T0212327

The Everyday Experiences of Reconstruction and Regeneration

Set within a wider British and international context of post-war reconstruction, *The Everyday Experiences of Reconstruction and Regeneration* focuses on such debates and experiences in Birmingham and Coventry as they recovered from Second World War bombings and post-war industrial collapse.

Including numerous images, Adams and Larkham explore the initial development of the post-Second World War reconstruction projects, which so substantially changed the face of the cities and provided radical new identities. Exploring these cities throughout the post-war period brings into sharp focus the duality of contemporary approaches to regeneration, which often criticise mid-twentieth century 'poorly conceived' planning and architectural projects for producing inhuman and unsympathetic schemes, while proposing exactly the type of large-scale regeneration that may potentially create similar issues in the future.

This book would be beneficial for academics and students of planning and urban design, particularly those with an interest in post-catastrophe or large-scale reconstruction projects within cities.

David Adams is a planner and geographer, and his current research focuses on questions of representation and experience of the urban realm. His research interests cut across the fields of urban theory, design and form. He is particularly interested in how urban townscapes are shaped and the (continued) impacts of post-Second World War reconstruction planning in Britain, with particular regard to data infrastructures, to encourage public participation in planning and development decisions. Elsewhere, his recent and ongoing projects include studies of 'guerrilla' gardening in its potential for planning discourses; and the reconfiguration of 'planned' cities through everyday practice.

Peter Larkham has a variety of research interests that cover different aspects of urban form. He has a long-standing interest in conservation, and in recent years has been focussing on the replanning and reconstruction of British towns after the Second World War. He has published numerous papers in his field and built a worldwide reputation. Peter has published over 65 refereed journal papers, written and edited several books and presented numerous papers at conferences in the UK and worldwide. His recent book is *The Blitz and its Legacy* (edited with Mark Clapson, in 2013) and three more books on post-war reconstruction and on urban form published in 2013–2014.

The Everyday Experiences of Reconstruction and Regeneration

From Vision to Reality in Birmingham and Coventry

David Adams and Peter Larkham

Routledge
Taylor & Francis Group

LONDON AND NEW YORK

First published 2019
by Routledge
2 Park Square, Milton Park, Abingdon, Oxon OX14 4RN

and by Routledge
605 Third Avenue, New York, NY 10017

First issued in paperback 2022

Routledge is an imprint of the Taylor & Francis Group, an informa business

© 2019 David Adams and Peter Larkham

The right of David Adams and Peter Larkham to be identified as authors of this work has been asserted by them in accordance with sections 77 and 78 of the Copyright, Designs and Patents Act 1988.

All rights reserved. No part of this book may be reprinted or reproduced or utilised in any form or by any electronic, mechanical, or other means, now known or hereafter invented, including photocopying and recording, or in any information storage or retrieval system, without permission in writing from the publishers.

Trademark notice: Product or corporate names may be trademarks or registered trademarks, and are used only for identification and explanation without intent to infringe.

Publisher's Note
The publisher has gone to great lengths to ensure the quality of this reprint but points out that some imperfections in the original copies may be apparent.

British Library Cataloguing-in-Publication Data
A catalogue record for this book is available from the British Library

Library of Congress Cataloging-in-Publication Data
Names: Adams, David, 1980– author. | Larkham, P. J. (Peter J.), 1960– author.
Title: The everyday experiences of reconstruction and regeneration : from vision to reality in Birmingham and coventry / David Adams and Peter Larkham.
Description: Abingdon, Oxon ; New York, NY : Routledge, 2019. | Includes bibliographical references and index.
Identifiers: LCCN 2018058373 | ISBN 9781472471178 (hbk) | ISBN 9781315558424 (ebk)
Subjects: LCSH: City planning—England—Birmingham. | City planning—England—Coventry. | World War, 1939–1945—Economic aspects—England—Birmingham. | World War, 1939–1945—Economic aspects—England—Coventry.
Classification: LCC HT169.G72 B54633 2019 | DDC 307.1/2160942496—dc23
LC record available at https://lccn.loc.gov/2018058373

ISBN 13: 978-1-4724-7117-8 (hbk)
ISBN 13: 978-1-03-240158-4 (pbk)
ISBN 13: 978-1-315-55842-4 (ebk)

DOI: 10.4324/9781315558424

Typeset in Sabon
by Apex CoVantage, LLC

Contents

Acknowledgements

Oral history data collection, as noted by leading figures working in this area, is an arduous undertaking, and we are deeply indebted to have access to the original interview transcripts from Dr. Lucy Faire's work on Coventry carried out in 2001. Thanks also go to Professor Phil Hubbard and Professor Keith Lilley for their support. Professor Carl Chinn also helped identify suitable respondents to talk to for the Birmingham leg of the research. Rob Orland, who maintains the *Historic Coventry* website, was also very helpful in providing contacts for possible interviewees. Thanks also go to the Library of Birmingham archives, the Birmingham Museum Collections service, and to the staff at Coventry's Herbert Art Gallery archives, particularly regarding material relating to the City Council's minutes. Of course, alongside James Roberts, Brian Redknap and the late John Madin, it should go without saying that the authors are indebted to all those people from Birmingham and Coventry and representatives from the cities' local history societies/groups, who reported their time and post-war experiences. Finally, we would like to thank Aoife McGrath at Taylor Francis for her help and patience with producing the manuscript.

1 The process and product of reconstruction

Introduction

Since the beginning of the twentieth century, social campaigners, prominent authors, urban reformers and politicians across Europe and the US have spoken of their concern that the large cities that had arisen as a direct and indirect consequence of rapid and sustained industrialisation were, in many respects, at the core of many of the social iniquities of contemporary society. There are rich amounts of literature describing the forces of late-nineteenth and early-twentieth century modernisation, urbanisation and industrialisation and their impacts on inhabitants' everyday lives. Although sociologists such as Durkheim (1893 [1997]) talked of the emancipatory potential of large industrialised cities, a more common prognosis was that large and expanding cities were places that led to a sense of personal anonymity and social impoverishment. There were concerns surrounding containing the growth of the large-scale cities and expanding provincial towns. For example, from 1898, Ebenezer Howard (1898) was developing the idea of the 'garden city', although very few were built, and there were more efforts on the scale of the 'garden suburb'.

In the 1920s, Lewis Mumford and Clarence Stein dreamed of a more spacious and commodious regional North American city, which would offer an attractive alternative to the outmoded lifestyle perceived to be associated with the densely populated compact city (Miller, 1989, page 199). In Germany, the idea of *Stadtlandschaft* was used as a way of fusing beneficial aspects of the town and country through planning interventions (Diefendorf, 1989, 1990). By the 1930s, though, when confronted with the disillusionment of the apparent failure of large-scale urbanism and the prospect of an impending 'total war' wrought by technology and air power, some architects and planners embraced the prospect of war with a certain enthusiasm. Though this might represent a particularly callous and unpalatable perspective, at the beginning of the Second World War, plans for the complete eradication of 'outworn' urban areas were produced for Rotterdam, Manchester, Coventry, Hamburg and other cities in Western Europe. In short, there was an expectation that future devastation would create the opportunity to build the 'City of Tomorrow'.

The promise of reconstruction

Of course, the bombing of British towns and cities during the Second World War was perceived as devastating, although the civilian death rate was not perhaps as high as had been expected, but the damage to buildings and infrastructure was, in many cases, more dramatic. Houses, services, offices, warehouses, factories, churches and other public institutions were widely damaged (see Clapson and Larkham, 2013). Of course, there are 'official histories' charting the nature of bomb damage, response and reconstruction (see, for example, Cullingworth, 1975), a more wide-ranging and critical perspective has emerged in many disciplines since the 1980s. Calder (1991), for example, has challenged the prevailing representation of the British stoicism under attack and called into question the official line that the people of Britain pulled together in a persistent flurry of communitarian and nationalistic defiance from 1940 to 1945.

Hewitt's (1994) exploration of the experience of those inhabitants living through the wartime destruction is also instructive in bringing into question prevailing 'official' accounts of the Second World War (see also Titmus, 1950; Gardiner, 2010). By specifically drawing on and prioritising the personalised testimonies of women in significantly bombed cities in Germany, Japan and England – as being representative of able-bodied persons who witnessed the severe disruptions of everyday domestic life – Hewitt's work does much to suggest that personal narratives should be embedded into studies of marginalised people in 'oral geography'. Of course, the stories of the wartime survivors form a poignant body of information about the direct, dreadful and dramatic impacts of warfare on cities. In addition, 'total war' – a massive act of *planned* urban devastation in itself – served to accelerate ideas of modernist urban planning, to the extent that Hewitt (1983, page 278) has argued that the 'ghosts of the architects of urban bombing' still loom large over the streets of many cities.

The idea that a 'greater good' might result from wartime destruction ushered in a social consensus in favour of planning that, according to some scholarly commentators writing on this period of planning history, involved all social classes (and political parties) (Stevenson, 1986; Düwel and Gutschow, 2013). Cherry (1988), for example, argued that war damage gave the opportunity to rebuild and a new social psychology in wartime Britain provided the determination. Writing in the British context, Higgott (2000, page 151) points out that modernism became 'a kind of vernacular in the years following the end of the Second World War as 'modernist practices and modernist forms moved out of the experiment of pre-war radicals and into the policies and programmes of large city authorities'. Many British plan authors at the heart of driving forward this post-war future were primarily or solely architects or surveyors rather than town planners, although some had joint qualifications, and the most eminent had been president of the Town Planning Institute (Thomas Sharp, Patrick Abercrombie, William Davidge,

William Holford, Stanley Adshead, W. Dobson Chapman). A number were linked through studying or teaching at the UK's oldest academic department in the field, the Department of Civic Design at the University of Liverpool (Larkham and Lilley, 2001). Groups such as the Town and Country Planning Association, the Town Planning Institute and the Royal Institute of British Architects were also important bodies in arguing that urban planning was vital if the rebuilding of the war-torn urban fabric was to result in a 'Better Britain' (e.g. Larkham and Lilley, 2012). There were, however, heated disagreements over *which* profession should drive the rebuilding programme. In terms of British post-war planning, the period of 1944–1950 was a crucial time in the development of ideas and processes: reconstruction prompted central government action, through the short-lived Ministry of Town and Country Planning (Cullingworth, 1975). The Town and Country Planning Act of 1947 also specified the role and nature of 'development plans', which then led to a more 'technocentric' (Diefendorf, 1989, 1990) approach to plan-making, using major infrastructure to deliver development, particularly regarding managing traffic congestion. It also, radically, 'nationalised' the right to develop land, thereafter requiring planning permission.

This interpretation of planning's ascendency into the heart of the British government is closely allied with a more general assessment of the post-war years as being enthralled by the idea of embracing all things 'modern' (Cullingworth, 1975; Lowe, 1990; Hewison, 1995). A view still persists that during the post-war period, Britain entered an era of improvement, with existing social arrangements being dismantled as discourses emphasising the desirability of order, cleanliness, speed and efficiency came to the fore (Berman, 1983; Harvey, 1989). For example, several commentators suggest that the zenith of modernity was in the mid-twentieth century, a time when self-confident ideas about science and progress found expression in cities designed to the precepts of bureaucratic state-sponsored planning (Hall, 2014). As with other professionals, the generally accepted view was that planners represented the informed experts, though as Scott (1998) suggests, this faith in the 'planner' was tragically misplaced. In his view, the creation of highly ordered urban environments whose flows could be analysed, assessed and ordered 'from above' made little concession to the local histories and practices of those people who lived, worked or socialised in them.

Reconstruction planning has its enduring myths. It has now become commonplace in Britain to blame 'the planners' for much that was rebuilt; from the purportedly utilitarian high-rise council estates, to the unpopular modernist urban designs and concrete buildings that arose, like unattractive phoenixes, on the bombsites of cities, or on the green fields allocated for new town development and the planned decentralisation of urban populations (e.g. Davies, 1972; Dennis, 1972; Amery and Cruickshank, 1975; Esher, 1981; Prince of Wales, 1989). Although these instructive accounts add nuance to the concept of the production of social space, there is a sustained academic focus on the influence of the professional 'planning experts', their lives and ideas and where they have been so often presented as the sole author of the plans (Larkham, 2013). Although there is a strong tendency to identify a named 'planner', in reality even consultants' plans were team products and the majority of plans were drawn up by local authority in-house teams. According to Cherry's work exploring the pioneers of British planning (Cherry, 1981), Britain had, following the end of the war, become increasingly receptive to bold new plans for reconstruction that were published during the war and in its aftermath (see also Bullock, 2002). This was, perhaps, a 'golden age' of British town planning (Mandler, 1999, page 209). But this was short-lived; the bold, radical, new plans faded quickly after 1947 (Hasegawa, 1999), side-lined by harsh financial realities and the new post-1947 Act formula for Development Plans.

For others, though, planning had a surprisingly limited role in shaping post-war change, suggesting instead that efforts should also focus on how the built environment evolved within a market-driven model, rather than – as some authors have argued – simply following the planners' idealised and ordered visions (Flinn, 2012). For scarcely any of these visions were ever fully implemented. Düwel and Gutschow's (2013) beautifully illustrated collection of essays from historians and architects of various cultures provides an analysis of the ideas and convictions on which post-war urban planning was modelled. Here it is suggested that 'we must ask again and again: what expectations did the "professional experts" . . . have of the city – and for future society' (page 10). While these essays offer a thorough and detailed analysis of 'official' accounts, plans, maps and representations, others have outlined the dangers implicit in writing planning histories relying mainly on such sources. Many descriptions of the process of post-war urban planning arguably offer a distanced view detached from the complexities of the socio-economic context in which reconstruction unfurled, or, indeed, the everyday lives of those residents living, working and socialising in reconstructed cities (Platt, 2015). Surely, acknowledging and understanding the experiences of surviving the replanning and reconstruction are as valid as the experiences of surviving the blitz itself.

A closer reading suggests that the reality underpinning the redesign and rebuilding is far more nuanced, involving a multitude of actors – working at different spatial scales – including planners, property owners, developers and architects (Marriott, 1967; Tiratsoo, 1990; Flinn, 2012, 2013; Shapeley, 2012). Local corporations, planners and city engineers were involved in the efforts to zone cities, produce greener housing estates and improve the physical, economic and environmental conditions (Greenhalgh, 2017). More broadly, perhaps, Sandercock (1998) and Yiftachel (1998), for example, suggests that more attention ought to focus on the idea of appreciating the contemplative knowledge of certain marginalised groups and communities and an acknowledgement of planning's 'darker side' and the inglorious episodes of the discipline's past. The broader issue with the idea that everyday lives are affected, a sense of community submerged or lost, as a direct (or indirect) consequence of planning and modernisation has been widely explored. And there have been several responses to Sandercock's (1998) call, with strong evidence of community-based or 'bottom-up' planning approaches emerging in different contexts.

However, as Ward *et al.* (2011) note, biographical and chronological approaches used to explore episodes in planning history have done much to create the illusion of planning being a discipline and profession dominated by a white, middle-class, male and Western perspective removed from the every-day realities and the experiences of those living in 'planned-for' and reconstructed cities. As Ward *et al.* (2011) suggest, the careers of influential male planners such as Frederick Law Olmstead, Daniel Burnham, Patrick Geddes and Lewis Mumford have all been chronicled in a biographical fashion. Cherry (1981) provides an early and instructive biographical account of eight prominent and influential British town planners including Patrick Geddes and Patrick Abercrombie. Cherry and Penny (1986) have discussed the role and influence of William [later Lord] Holford (1907–1975) who succeeded Patrick Abercrombie as Professor of Civic Design at the University of Liverpool in 1937. He was an influential member of the Planning Technique section of the Ministry of Town and Country Planning. Autobiographies are rare, although Abercrombie's unpublished autobiographical typescript has recently been made available for study (see: https://eprints.utas.edu.au/23327/1/Abercrombie-memoirs.pdf), and other influences can be deduced from the work of those reconstruction planners who became academics (Johnson-Marshall, 1966). Those involved in national or local political decision-making may be more motivated to write autobiographical pieces justifying their actions (e.g. Reith, 1949; Price, 2002). The feelings and motives of civil servants, though, tend to remain obscure (though see Ward, 2012; Sheail, 1981, 2012; and both Sharp and Holford worked for the Ministry at times).

Autobiographical and ethnographical accounts, testimonies and letters to local newspapers – although these are all partial and probably biased accounts – have been studied to reveal something of the experiences of living through urban change (see, for example, Beaven and Griffiths, 1999). Despite these and other studies, however, oral history methods remain relatively under-employed in planning history. These narratives tend to belong to planners and architects rather than local people (Voldman, 1990; Gold, 1997, 2007), though such in-depth research using a life history approach often elicits information from actors that would not have otherwise been available via 'official' archival sources. A more 'grounded' approach has been employed by Llewellyn (2003, 2004) in his study of Kensal House, Ladbroke Grove and North Kensington, London, and attention is given to the narratives of residents and how they responded to Maxwell Fry's ideas for Kensal House and to the interplay between residents and other actors in development of the scheme. Hubbard *et al.* (2003a, 2003b, 2004), in the context of the reconstruction of the 'rationally planned' Coventry, and Adams (2011) and Adams and Larkham (2013) in relation to post-war Birmingham have employed a 'traditional' oral history approach with residents who lived through the post-war replanning and subsequent reconstruction of these two cities. These studies attempt to outline the dangers involved with constructing planning perspectives that are divorced from the diverse personal experiences of those living and working in reconstructed cities. This work is particularly useful from a methodological perspective as it brings people's memories of living in post-war Coventry and Birmingham into dialogue with existing published and unpublished accounts of the cities' redevelopment, resulting in a much richer, multi-layered account.

Although Hubbard *et al.* (2003a, 2003b, 2004) draw on oral history interviews of those residents who lived through the reconstruction of central Coventry, and Adams's (2011) similar work for the Birmingham's 'un-planned' city centre, these recollections recorded by some residents tend to represent a reaction against the intensification of modernity. Counter-memories, in this sense, contrast with the 'abstracted' design ambitions of planners, architects and other decision-makers who, some feel, 'impose' their insensitive planning ambitions onto the lived spaces of those residents living, working and socialising in certain urban spaces. On the other hand, however, bringing the narratives of those individuals who were implicated/involved in the process of modernising the urban built environment into juxtaposition with those residents who inhabited newly constructed spaces offers a fertile line of inquiry, digging deeper into the ways in which the built form is 'lived and performed' beyond 'traditional' forms of representation. Moreover, given that the post-war built form of many British cities is now becoming increasingly subject to radical alteration or demolition it should not only be recorded but its significance (in both a lay sense as well as the technical meaning used by official conservation decision-makers such as Historic England) needs to be appraised, especially given how the material and non-material sites can affect a community's or an individual's ongoing engagement with urban space. Coventry, for instance, has recently been chosen to be the UK City of Culture for 2021, seen by some, at least, as being an excellent chance to celebrate aspects of the city's post-war urban form, but also to help reverse the city's image of being in a permanent state of 'grimy decline'.[1]

Methodological note

One of the central aims of this study, therefore, was to try to capture the embodied understandings and personalised social histories of those who can remember the process of post-war change. Emphasising these perspectives, this research seeks to look beyond the 'planners' professional 'intellectual representation of space' (which may differ from that of other professions such as architects) in an effort to explore the way in which urban space is given meaning 'through people's physical gestures and movements' (Lefebvre, 1991, page 200). This helps us to understand the relationship that exists between people and the urban built form. Several researchers argue that walking maximises the potential for place-based narratives, as the environment through which we walk can prompt further discussion that may not occur in a typical sit-down interview (Kusenbach, 2003). One fundamental strength of this approach is that the material environment can act as a prompt for people's attitudes, feelings and recollections concerning locations in a variety of ways, some unexpected.

This research therefore used a 'go-along' approach, thus combining the benefits of a traditional oral history with the environmental prompts of walking. Before the 'go-alongs' with residents took place, the transcripts from Hubbard *et al.*'s (2003a, 2003b, 2004) earlier study of Coventry (43 individuals interviewed during 2001) and the data gathered by Adams (2011) for Birmingham (22 individuals interviewed during 2007–2008) were re-evaluated. They were manually coded, and broad themes were generated from this analysis. Following this analysis, a 'go-along' study was conducted in central Coventry and Birmingham. Interviewees were recruited through an appeal to local history groups. Respondents came forward from several such groups. Nine individuals were recruited for the Coventry leg of the research – (1 female and 8 males, the oldest being 81). Thirteen people responded to the call from Birmingham (9 females and 4 males, the oldest being 80). The average age of respondents for Birmingham was lower than in Coventry (70 compared to 76). All 'in-the-field' interviews were conducted during 2012. On three occasions, a friend and fellow interviewee accompanied an interviewee.

This research tended to focus (though not exclusively) on the broad theme of post-war urban change and how people's spatial routines (taking account of their place of residence in the city, where they worked, how they travelled to work, shopping expeditions and trips to the cinema) altered during this period. Participants were therefore encouraged, while walking, to weave together their general experiences of the post-war rebuilding activity with their everyday memories of shopping, commuting to work and other interactions with the changing urban environment, alongside other life events such as marriage, moving to a new house or having children. The interviews offered a space to conduct a reflective analysis. They covered issues relating to an individual's thoughts about the changing urban landscape during the 1950s, 1960s and early 1970s alongside their experiences/feelings towards the current city centres. The research focussed on the opinions that people expressed about their shifting spatial practices and routines (how they got to work, how they went shopping and how they spent their leisure time). Questions were typically framed using an open-ended format, such as: 'Tell me about your experience of walking through the Upper Precinct (Coventry) after it had been constructed'. Other questions were broader, including: 'Tell me about your feelings towards the rebuilding of Birmingham – did this alter your routines in any way?' Ultimately, the histories of residents as practice-based and place-specific narratives allowed for a richer understanding of post-war reconstruction.

Variables such as age, general length of residence (and type of occupation) were also recorded. While walking around both city centres, respondents' points tended to broaden out and develop into a much wider discussion surrounding the role of planning interventions during and after the end of the Second World War, as well as their thoughts and opinions on more recent attempts to redevelop the city centres. The duration of the go-alongs differed according to the availability of the respondent: occasionally these would last for a rushed 30 minutes when a respondent wanted to take time out during their shopping excursion. At other times, interviews could last several hours, with the length of the interview entirely determined by the respondent. On average, the interviews lasted for 70 minutes: the longest lasting for over two-and-a-half hours. Individuals were occasionally prompted to offer an opinion on certain features situated in the city centres, focussing on specific buildings, pieces of public art or street furniture. For the purposes of this study, participants gave permission for their interviews to be cited anonymously; where this research cites respondents' narratives, only their first name and last initial, date of birth and date/style of interview has been used (e.g. John A., born 1937, go-along).

It is also openly acknowledged that the sample of respondents was not necessarily representative of the two cities' population during these years, and there was also a natural initial concern that the call for respondents would have attracted more confident middle-class professionals, although this proved not to be the case. Though this does not constitute a representative sample of those who lived and worked in the city during the years of reconstruction, the range of backgrounds from which the respondents were recruited gave little reason to suspect that the sample was biased towards any one socio-economic group.

As Gold (1997, 2007) points out, seeking the views of those 'actors' involved with the design, implementation, scale and overall context in which post-war buildings and infrastructure were shaped offers a valuable insight into the production and consumption of post-war planning ideas. These perspectives, whether contemporary or collected more recently, do help to expand on the actors' rationales given in other commonly used sources such as committee minutes and promotional material. To this end, therefore, alongside the personalised narratives of residents, the study also draws on a small selection of interview material from selected 'actors'. These were taken from several sources, including the National Sound Archive (NSA) in London. The NSA offers some valuable material on the lives and work of significant twentieth-century architects compiled by skilled interviewers (Gold, 2007). Together with Donald Gibson, Coventry City Architect – identified as being instrumental in shaping the wartime and post-war plans for the city's reconstruction – the NSA included other notable personalities that help shape the reconstruction of post-war Coventry. Unfortunately, few, if any, of these actors centrally involved with the department still survive. However, this research did make use of the previously conducted interviews with Donald Gibson and Percy Johnson-Marshall, which were acquired from the NSA and re-evaluated.

Anthony Sutcliffe's (1967–1969) detailed interviews with prominent people closely involved with the post-war reconstruction of Birmingham also proved useful. These include interviews with Herbert Manzoni (the City Engineer and Surveyor from 1935–1963, d. 1972), Sheppard Fidler (the City Architect from 1952–1962, d. 1990), the influential local councillor

(Sir) Frank Price (d. 2018) and locally trained architect John Madin (d. 2012). The transcripts of these interviews were also re-assessed. Several other powerful personalities helped shape Birmingham's post-war development, including the Birmingham-born and educated Jack Cotton (d. 1964) who had a hand in several prominent post-war buildings in the city centre. As with Coventry, however, those 'actors' closest to the reconstruction are either no longer alive or untraceable, though the study also made use of autobiographies of these figures where these are available (e.g. Price, 2002).

Although a wide range of locally trained architects were active in Birmingham during and after the war, two are significant, as their buildings dominated certain quarters of the city, James Roberts and John Madin. Both are widely regarded as having a profound influence on the reshaping of the city. They were responsible for the design of three of the city's most significant examples of modernist post-war architecture: the Smallbrook Ringway, the Rotunda and the Central Library. This project also interviewed James Roberts and John Madin while documenting and interpreting their work using the Birmingham Central Library archive and other sources. Brian Redknap, the City Engineer for Coventry from 1974 to 1989 and author of *Engineering the Development of Coventry* (published in 2004), was also interviewed using a more traditional semi-structured approach adopted by Gold (2007).

Unlike Sutcliffe's earlier work on Birmingham, the interviews with Roberts, Madin and Redknap allowed for more reflective narratives to be constructed. The additional four decades' distance from the events and personalities probably facilitated this more reflective response. Discussions typically focussed on their early lives, their academic experiences, major career episodes, their involvement with different developers in the context of the reconstruction of Birmingham and Coventry, their relationships with the respective City Council officers/departments during the post-war period, their feelings towards the resultant built form and their reactions to the changes that have taken place in more recent years. Madin's buildings are suffering disproportionately from demolition. These stories were also particularly useful because they were captured at a time when the reconstruction-era commercial heart of both cities was being significantly changed, with refurbishment and demolition all having an increasing effect. Gold openly admitted that the architect interview is 'wholly unreliable' and that it tends to promote a particularly individualist and visionary perspective (Gold, 1997, page xxii). The interviews with Roberts and Madin were certainly coloured by their desires to colour the past – they were not altogether dispassionate and value-neutral. However, Gold also recognised that oral sources have the potential to explore how the post-war transformation of British cities rested on a coalition of other factors (and agents) in the design and development process (Gold, 2007, page 14).

As well as talking to Roberts, Madin and Redknap and local people who remembered Birmingham and Coventry in the post-war years, this study draws on contemporary written accounts, including the Ministry of Housing and Local Government files held in the National Archives, Council minutes, other material in local archives and, for Coventry, Percy Johnson-Marshall's archive held at the University of Edinburgh. Contemporary professional journal material is also invaluable. Other accounts were drawn from articles published in national daily newspapers and letters written to local newspapers, especially *The Birmingham Evening Mail* and the *Coventry Evening Telegraph*. Although it is also acknowledged that these sources are no less problematic than oral history interviews in many ways. Letters sent to local newspapers were undoubtedly 'vetted' by the editorial team, and of those letters that were printed, there is little indication of the sender's age, gender or social status in most instances. Notwithstanding these issues, an analysis of local newspapers can be instructive in exploring local reactions to planning proposals.

Drawing on and using people's memories is not without its limitations; earlier critiques of oral history approaches have focussed (in the main) on issues concerning validity and reliability and whether such a method could be interpreted as being an accurate representation of a person's past. Any approach designed to 'bring memories out' is therefore a complex and creative process of story reconstruction. Although an element of a person's original story will re-appear when being interviewed, the remembered account is almost never the same – it is being re-created through the process of remembering. Taken in this way, therefore, one of the key strengths of exploring the everyday experiencing of post-war reconstruction is that people's recollections are being re-written 'through memory'. Rather than being a limitation, however, this research, benefitted strongly from the ability of the interviewees to reflect on the past from the perspective of the present. Of course, there is an obvious bias to memory: some events, historical figures, specific buildings or places are remembered, whereas others will be hidden from view, distorted or consciously (or even sub-consciously) re-arranged to produce a more palatable narrative for the listener/interpreter. Despite this, however, that there is much value in exploring the personal memories that are necessarily infused by experiences of the local environment, whilst being mindful that people's recollections are not independent of the experience of such as events that have taken place at national or international scale.

Notwithstanding these warnings, 'official' accounts of post-war reconstruction have been – and continue to be – produced for two cities in the British midlands, Coventry and Birmingham. For example, the post-Second World War planning and reconstruction of the Coventry's city centre has received extensive critical consideration with notable studies. In terms of its significance to post-war reconstruction, Coventry very quickly set about the replanning of the centre, encouraged by the proclamations of central government (in particular then Minister, the forceful character Sir John Reith), to *plan boldly* and was considered, initially at least, as the forerunner test-case reconstruction project. Led by Donald Gibson, the pioneering City Architect, the scale and comprehensive nature of the replanning ensures that Coventry lies somewhere between Plymouth, with the realisation of Patrick Abercrombie's grand, formal plan and Thomas Sharp's more sensitively handled replanning of Exeter (Gould and Gould, 2016, page 103). In contrast, nearby Birmingham (some 30-km north-west of Coventry), guided by Herbert Manzoni (later Sir Herbert), the City Engineer and Surveyor, a powerful figure nationally and locally in his profession, did not produce an overall city-wide official plan. Yet the city also represents a different example of major post-war reconstruction in the UK – this intense activity produced the reconstruction of large sections of the bombed city centre, an inner ring road necessitating a private Act of Parliament and the continuation of large-scale slum clearance and rebuilding begun before the war. Like Coventry, much has been written of the city's approach to post-war reconstruction, including an official history (see Borg, 1973; Sutcliffe and Smith, 1974).

The narratives elicited from respondents, of course, focus on different facets of the urban landscape and their readings of it reflect past narratives and present intentions. This book focuses particularly, although not exclusively, on the city centres within and adjacent to the both cities' inner ring road (Figures 1.1 and 1.2).

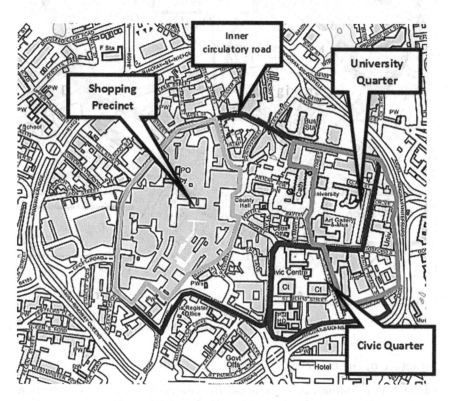

Figure 1.1 Coventry city centre. This shows the inner ring road, the inner circulatory road, the Shopping Precinct, University Quarter and the Civic Quarter

Source: Crown copyright reserved

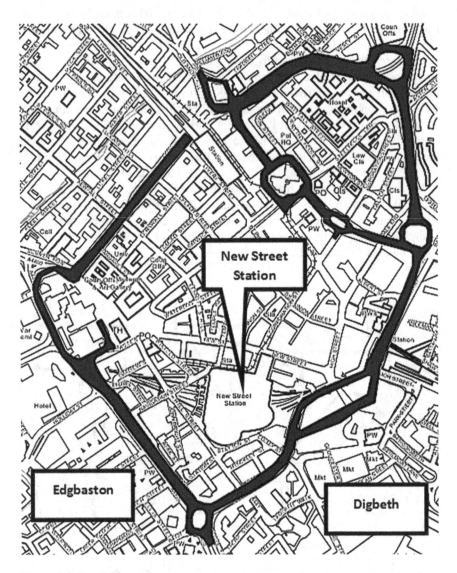

Figure 1.2 Birmingham city centre
Source: Crown copyright reserved

Structure of the book

Chapter Two outlines in some detail the context in which reconstruction planning in Britain was born by reviewing the professional, political and economic context in which wartime and post-war plans were produced, before going on to shift emphasis to focus on the reception and consumption of this replanning activity. Chapter Three re-explores the evolution of plans for the redevelopment of Coventry and Birmingham through the 1940s to the 1970s, when ideas for the cities' reconstruction developed and when the cities experienced their most intense period of rebuilding. Attention then turns to the personalised narratives in Chapter Four with residents to elucidate citizens' experiences of war and how they adjusted to their new city centres. As with other cities, both Coventry and Birmingham are again being physically and symbolically re-assembled to create a city centre in mood with the times. These changes are discussed in Chapter Five. Reflecting the ethos of regeneration, both realised and proposed redevelopments have been predicated on the disavowing of the 'unruly' (cf. Ashworth and Tunbridge, 1996) elements of the post-war urban landscape. As the redevelopment of much of the built form from the reconstruction period is now proceeding apace, Chapter Six reviews elements of the rebuilding and considers how personalised narratives can be used to inform debates surrounding the conservation of the built form associated with this significant period. Chapter Seven concludes by setting Birmingham and Coventry in a wider context by positioning the post-war planning ideas associated with these two cities in relation to other British cases.

Note

1 See www.bbc.co.uk/news/entertainment-arts-42272675.

2   Designing and delivering reconstruction

Introduction

The decade of the 1940s was a landmark period in British planning for several reasons. These included the swift reaction to the impact of wartime bomb damage and the clearance of bomb-damaged properties; the production of new, bold reconstruction-style urban plans, some with radical and modern design approaches; new planning legislation including the 1944 Town and Country Planning Act, commonly called the 'Blitz and Blight' Act and the 1947 Town and Country Planning Act and the introduction of new facets of planning including New Towns, Green Belts, National Parks and Listed Buildings. But the genesis of these responses lies firmly in the interwar period and even before the First World War. It would be a mistake, therefore, to assume that planning activity during and after the Second World War was inherently 'new' in Britain or elsewhere (see Düwel and Gutschow, 2013). As Larkham (2013, page vii) argues, perhaps the major contribution of the 1940s was that British planning became a more extensive function: a mainstream and central feature of central and local government. The key legislation was significant for over half a century and is still influential two decades later. By any measure, that is a noteworthy achievement.

Important planning issues had arisen or, more accurately perhaps, been acknowledged, a decade or more before the 1940s. These included the need to provide more and better housing, both for a changing population and because much of the inherited housing stock of the Victorian, and even late-Georgian, period was identified as substandard slums in need of wholesale clearance (Ward, 2004). Increasing domestic (and international) political pressures in the 1930s led to renewed demands for new forms of housing and its attendant industrial employment and other services away from the existing over-crowded urban centres. At the same time, vehicle ownership was increasing rapidly, and this had obvious implications for existing street layouts and urban forms inherited from previous centuries, repeatedly resulting in increasing levels of congestion and traffic-related pedestrian fatalities (Tripp, 1938, 1942; MoWT, 1946). These issues connected to other societal changes: for example, increases in paid leisure time led to pressure on facilities, generating the need for more recreational amenities and green spaces, both urban and rural and recognition of the significance of historic places and tourism (Hewison, 1995). The British planning system, essentially un-coordinated as it was up to the end of the 1930s, was largely unfit to respond appropriately to these changing circumstances and challenges.

Following the outbreak of war in September 1939, the immediate fear was of aerial attack, and preparations for air raids were stepped up. The war was undoubtedly a deeply disturbing experience for those individuals and communities that were soon subject to the aerial bombardment. Concentrated attacks from late 1940 into 1941 resulted in a badly damaged London (the 'Blitz') and other industrial towns (Merseyside, Birmingham, Plymouth, Coventry and others). In 1942, a series of smaller raids – the *Baedeker* raids – specifically targeted at historic cities including Bath, Canterbury, Exeter, Norwich and York, were a deliberate attempt to damage civilian morale (Rothnie, 1992). Throughout the war, small raids – sometimes even just a lone aircraft – contributed damage to property and stress to people. In 1944 and 1945, the V-weapons created more damage, particularly in the southeast of England. By the cessation of hostilities, an estimated 71,000 tons of bombs (and V-weapons) had fallen on the UK (Titmuss, 1950). In mid-1946 the Ministry of Health (which still maintained some responsibility for some planning issues, especially housing-related) estimated for the UK that some 22,000 houses were destroyed/damaged beyond repair; 4,698,000 damaged in some way,[1] and cumulatively, 'a total of 3,745,000 different houses in the United Kingdom were either damaged or destroyed during the Second World War' (Titmuss, 1950, pages 329–330).

The professional planning perspective, even as the bombs rained down on Britain, was that the wartime damage provided an 'opportunity'. Although those who lost relatives, friends and homes surely did not see it in this light, this word was writ large in many contemporary publications, both professional and mass market and in radio programmes relating to the prospect of post-war reconstruction (see, for example, Tubbs, 1942; Ashworth, 1954). The planning response was to produce numerous local, city-wide and regional plans, and over 250 'plans' of various types have been identified between 1940 and *c.* 1952 (Larkham and Lilley, 2001). Planning, as with other areas of professional activity, became an accepted part of public discourse. Post-war ideas of 'replanning' and 'reconstruction' expanded to issues far outside the physical rebuilding of bomb-damaged urban areas – society and the economy, among other things, were to be reconstructed (Kynaston, 2015).

Many of the plans, particularly those produced early in the period, were visionary, even radical, drawing on modern approaches to the replanning 'problems' (Hasegawa, 1992). This involved the substantial reconstruction of towns, whether they had suffered significant damage or not. In fact, many little-damaged or undamaged towns jumped on the replanning bandwagon, afraid of otherwise being left behind in the new post-war restructured urban hierarchy. New 'planned for' infrastructure, particularly ring roads, provided the armature for this urban restructuring, and new technology and materials, especially system building, produced new architectural and urban forms (Bullock, 2002). These sometimes-innovative visions were made accessible to the consuming public principally in two ways: first, through numerous staged exhibitions, many of which attracted substantial proportions of the population of the town concerned (Larkham and Lilley, 2012). Second was the publication of, occasionally lavishly produced, plans or extracts, in forms varying from free or cheap booklets/pamphlets to hardback books costing several pounds. Although these representations often veered close to being an exercise in public 'propaganda', the relationship between effective public consultation and people's views being successfully integrated into plan-making is far from clear (see Larkham and Lilley, 2012) – few of the concerns that were raised by these presentations were incorporated in final plans or implementation. This chapter provides an overview of the British wartime and immediate post-war reconstruction plans, their professional, political and economic context and lastly, the reception and consumption of this replanning activity.

Pre-war concerns

Reflecting on the sprawling suburbanisation of the interwar years, especially around London, there were attendant calls for planning to play a more prominent role in shaping the built environment. There was widespread dissatisfaction with housing and ways of living – at least among the educated elite. In his pre-war novel, *Coming Up for Air*, George Orwell was less than sanguine over the problem of having 'long, long rows of semidetached (suburban) houses', of 'stucco front, creosoted gate, privet hedge' (Orwell,1939; see also Harris and Larkham, 1999). Following his experiences of fighting in the Spanish Civil War, he wrote of the seemingly peaceful suburbia lying outside London – this was an England insulated from the wartime threat emerging on the continent (Orwell, 1938) which, of course, included aerial bombing.

Thomas Sharp (1936 [1938], page 58), then a university lecturer in planning and architecture, complained bitterly of how the 'Midlands and the North advanced grimly along the hideous road of industrialism', while the South . . . slumbered fitfully in tree-folded parks and half-dead small towns'. Sharp's 'brilliantly argued' (Ward, 2004, page 86) Penguin paperback, *Town Planning* (1940), was a best-seller even in wartime, selling tens of thousands of copies despite paper rationing (Stansfield, 1981) and did much to encapsulate the problems facing planning and make them accessible to a much wider public. In a similar way, E.M. Forster (1937, page 45) scathingly commented on the issues facing interwar Britain: 'there seems no hope of checking the general destruction . . . the population increases, the means of transport increase, the needs of the fighting services are allowed to increase. Something has to decrease, and it has to be the woods and downs, hedges and birds'. For Thomas Sharp, there were three options available: 'we can continue to live in stale and shameful slum towns. Or in sterile and disorderly suburbs. Or we can build clean proud towns of living and light. The choice is our own' (Sharp, 1940, page 114). Two years later and within a Birmingham context, *When We Build Again* (a book and film focussing on Bournville Village in Birmingham and funded by the Cadbury's, a firm and family with a long interest in planning) presented a visual interpretation and embodied some of Sharp's thoughts arguing vehemently that there 'should be no uncontrolled building, no more ugly houses and straggling roads, no stinting of effort before we build again' (Bournville Village Trust, 1941).

Given the levels of high unemployment, precipitated by the Wall Street Crash and the 'Great Depression', urban congestion, a rise in private car ownership (Gold and Ward, 1994)[2] and the sprawl of suburbia, the then government appointed a Royal Commission – which included the input of renowned architect and planning consultant Professor Leslie Patrick Abercrombie – to examine the 'distribution of the industrial population'. Abercrombie was an influential figure in national UK planning debates and garden-city initiatives and his influence are widely discussed elsewhere (Dix, 1981; Cherry, 1989; Higgott, 1991–1992; Hall, 1995; Jones, 1998; Lambert, 2000; Essex and Brayshay, 2005). Influenced by Patrick Geddes' approach of 'survey-analysis-plan', his distinguished planning career began to blossom in 1915 when he replaced Stanley Davenport Adshead as Lever Professor of Civic Design at the University of Liverpool (Dix, 1981). Abercrombie was responsible for nine reconstruction plans; only Thomas Sharp with his plans for Durham, Exeter, Oxford, Salisbury and Taunton, among others, produced more (with eleven). Writing of Patrick Abercrombie's influence, the *Architect & Building News* reported loftily that 'God re-made Britain from the designs of Professor Abercrombie' (*Architect & Building News*, 1943, page 22). The Commission's conclusions, outlined in the Barlow Report of 1940, were that massive concentrations of populations were economically, socially and strategically undesirable, and it recommended the establishment of a central planning authority charged with dealing with the regional balance of in the distribution of employment, planned metropolitan decentralisation (Ward, 2004).

Reconstruction is in the air

Professionals, primarily middle-class reformers originating from the centre-left, generated the *desiderata* of reconstruction (Ward, 2004). Wartime conditions did make many more people become more dependent on state provision of services (Kynaston, 2007) and these conditions and military service, led more people to be more used to some form of centralised control (in this case, state intervention). As A.J.P. Taylor (1961) later pointed out, the *Luftwaffe* was a powerful missionary force in generating public support for a welfare state and the sense of different groups 'coming together' to fight on the home front. Marwick (1968), however, keenly noted that the effect of evacuation of children to rural (and semi-rural) locations, for example, was that more often middle-class families were confirmed in their prejudices about the dirty fecklessness of the working classes (see also Wheatcroft, 2013). Nevertheless, universal provision of welfare services was a hallmark feature of one of the most influential reports to come out of the war, the Beveridge Report of 1942. The report's publication received a rather cool reception from the Cabinet. Churchill commented a month after the report's publication that 'ministers should, in my view, be careful not to raise false hopes, as was done last time by speeches about "homes for heroes"'. He urged caution against any loftier ambitions of achieving 'visions of Utopia or Eldorado' and 'promises of the future' (Churchill, cited in Brooke, 1992, pages 171–172).[3]

Nevertheless, evoking an evangelical spirit, the rhetoric of the Beveridge Report was of the struggle against the 'five giants on the road to reconstruction', Want, Ignorance, Squalor, Idleness and Disease (Beveridge, 1942). There are arguments to suggest that the creation of the Coalition government (formed in May 1940) was both an obstruction to progress towards post-war rebuilding *and* an opportunity for Labour ministers to exercise a certain level of influence in shaping plans for post-war reconstruction (see, for example, Malpass, 2003). Following the fall of France in June 1940, there were calls for more morale boosting reforms, which led to Churchill appointing a committee responsible for outlining plans for reconstruction (Cullingworth, 1975). As early as October 1941, John Reith was appointed to lead the government's response to reconstruction; this came largely in the form of the short-lived Consultative Panel on Physical Reconstruction.[4] Initially, under the auspices of Reith,[5] then Lord Portal (1942–1943) and finally William Morrison (1943–1945), the Panel gave a national sense of direction to urban planning (Flinn, 2013).

Under Morrison's leadership, an advisory group – the Advisory Panel on Redevelopment of City Centres – was formed in May 1943 (National Archives [hereafter NA] HLG, 88 / 8, 1943; Hasegawa, 1992). The need for a government response to reconstruction was underscored with Beveridge's opening of a notable public exhibition, *Rebuilding Britain*, at London's National Gallery in 1943 (*Architectural Review*, 1943). 'How can the war on Squalor be won?' asked the accompanying catalogue, referring to the five evil behemoths he hoped to eradicate. The answer, Kynaston (2007, page 31) notes, was to 'plan on a national scale'. The publication of the Scott Report (in 1942), which dealt ostensibly with tighter planning controls to preserve the British countryside (and its productivity), together with the Uthwatt Report (also published in 1942) concerning 'betterment and compensation' associated with appreciating land values, set the agenda for the government's commitment to building a better post-war world (Ward, 2004).

The Advisory Panel spent much of 1944 surveying several British cities selected as examples for post-war reconstruction. The technical planning section of the Panel began work on an advisory publication for local authorities; though most of this work was carried out during wartime, the Ministry of Town and County Planning publication, *The Redevelopment of Central Areas* did not appear until 1947 (MTCP, 1947) – perhaps not coincidentally the date of the major new planning Act. By then many 'advisory outline' reconstruction plans for central areas had already been written and widely publicised. Furthermore, the 1944 'Blitz and Blight' Act gave local authorities extended powers for reconstruction, much to the chagrin of some Conservatives and landed interests (Hasegawa, 1992). While Abercrombie's *Greater London Plan* and Abercrombie and Forshaw's *County of London Plan* (Forshaw and Abercrombie, 1943; Abercrombie, 1945) for the reconstruction of heavily bomb-damaged London essentially set the agenda for post-war planning, both in process and at the regional/city-region scale (Ward, 2004).

Plans to guide urban change

Throughout Britain, from the period from (around) 1940 through to the early 1950s, a set of reports and published plans were commissioned by public bodies – and a small number of other organisations and individuals – that imagined a vision of a progressive urban future, one that promised a powerful technocratic and buoyant set of schemes for the reconstruction of British cities and towns (Perkins and Dodge, 2012). These plans and reports wove together a distinctive narrative of urban change. At least in the early part of the decade, there was a broad consensus in favour of the activity of planning (Addison, 1975; Stevenson, 1986; Cowan, 2013). Most of these plans were written before the inception of the momentous 1947 Town and Country Planning Act, which ushered in a much more formalised and prescriptive textual approach towards the conception of planning frameworks, to be expressed as 'Development Plans' (Larkham, 2013). These early documents were largely unconstrained by legislation and, as a result, their content was perhaps more directly reflective of the ideas and beliefs of a small number of planners, architects and engineers. Although there were tensions between established professionals (architects, surveyors and municipal engineers) about who was responsible for leading the new 'planning' activity, many of the plans were much better illustrated, communicated ideas more clearly and were much more widely and positively reviewed in the professional and lay press than the post-1947 Act's Development Plans.

Early historiographies of British post-war planning have suggested that the publication of wartime and post-war plans represented an opportunity seized by an embryonic strain of local 'planners' to 'improve' their towns and cities (Ashworth, 1954). Hasegawa's (1992, 1999) informative work argues that the ideas surrounding reconstruction presented a unique opportunity for planning to engender a sense of greater social fairness within a society dominated by class divisions and inequality. Indeed, key individuals and groups (notably the Town and Country Planning Association, the Town

Planning Institute and the Royal Institute of British Architects) argued that urban planning was vital if the rebuilding of the war-torn urban fabric was to result in a 'Better Britain'. BBC radio talks and publications such as the Town and Country Planning Association's *Rebuilding Britain* series (TCPA, 1943, the *Planning and Reconstruction Year Book* (Osborn, 1942) and exhibitions organised by the Royal Institute of British Architects (RIBA) (Figure 2.1) helped build a consensus around the idea that rebuilding was

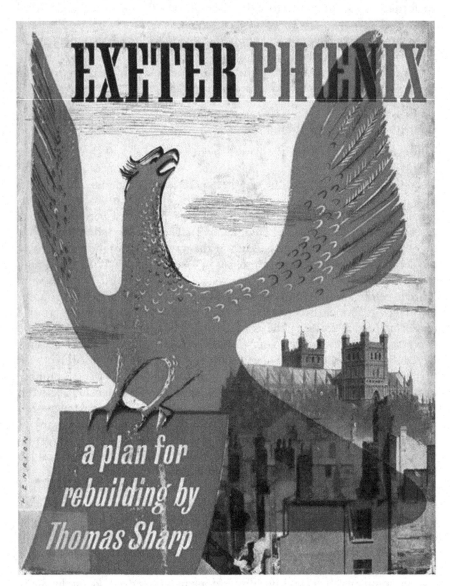

Figure 2.1 Front cover of Thomas Sharp's 1946 *Exeter Phoenix*
Source: Reproduced with permission of John Pendlebury

inevitable (Gold and Ward, 1994), and this included undamaged as well as bomb-damaged areas.

The idea that a greater good might result from wartime destruction thus ushered in a social consensus in favour of planning that, according to some scholarly commentators writing on this period of planning history, involved all social classes (and political parties) (Stevenson, 1986). This interpretation of planning's ascendancy into the heart of British government is closely associated with a more general assessment of the post-war years as being enthralled by the idea of embracing all things 'Modern' (Cullingworth, 1975; Lowe, 1990). A view persists that with the post-war period Britain entered an era of 'modern' improvement, with existing social arrangements being dismantled as discourses emphasising the desirability of order, cleanliness, speed and efficiency came to the fore (Harvey, 1989). This is exemplified by a comparison of contemporary planning to medicine and treatment of the body (Bartram and Shobrook, 2001).

Many commentators suggest that the zenith of modernity was in the mid-20th century, a time when 'muscular', self-confident ideas about science and progress found expression in cities designed to the precepts of bureaucratic state-sponsored planning (for example, Berman, 1983; Harvey, 1989; Hall and Tewdwr-Jones, 2010). Ideas of breaking away from the vestiges of the past were very much apparent in Lewis Mumford's *Culture of Cities* (1938), and it is known that his ideas directly influenced some reconstruction planners. Mumford argued that a move away from the 'ghastly, relentlessly cumulative error leading to social corruption and impotence' (Checkland, cited in Dyos, 1968, page 346) that characterised the organic growth of the 'modern' town, was tempered by the promise of a resurrection through effective city planning – verdant parkways, community-focused neighbourhood units, planned decentralisation and garden cities.

The influential Modernist architect-planner, Maxwell Fry, asserted that the early wartime bombing raids had revealed to 'the British public' how disorderly 'our cities . . . are! After experiencing the shock of familiar buildings disembowelled before our eyes – like an all too real surrealism – we find the cleared and cleaned up spaces a relief . . . we have a hope for the future, opportunities to be taken or lost' (Fry, 1941, pages 16–19). Similar messages emerge in the final quarter of Arthur Korn's (1953) *History Builds the Town*, which presents a 'particularly ruthless' (Hebbert and Sonne, 2006, page 18) interpretation of these ideas. As a communist representative for the proletarian leanings of CIAM IV (*Congres Internationaux d'Architecture Moderne*), and a key driving force behind the British-based MARS (Modern Architectural Research Group) plan for London, Korn's clarion call was for the erasure of the existing urban fabric to make way for the capital to be re-planned/re-structured around welfare-based principles (Gold, 1997, 2007).

Similar themes are apparent in the Finnish architect Eero Saarinen's (1943) juxtaposition of the 'urban disease' associated with the congested and sprawling settlements of the twentieth century and his prescriptions for the cleansing and reorganisation of cities 'of the future' in order to grapple with prescient issues of traffic management. In the British context, this represented a 'golden age' of planning: a period when, although the profession was youthful, planning could play a significant part in shaping human activity (Ward, 2004). Maxwell Fry aptly captured this attitude in 1941, writing at the height of the bombing:

> Given the will to plan, we could in a quarter of a century or less, substantially transform our worst towns. Where they are black with soot, they could be at least partly green with trees and grass. . . . It should be part of our national genius to do this sort of thing well, and if we do, invention will flow again.
>
> (Fry, 1941, page 20)

As with other professionals, planners were the informed experts and the subject was perceived as technical more than artistic – Abercrombie called planners 'technicians'. There have been several explorations of how the plans were the result of the conceptual work of the 'great' and 'visionary' master planners (see, for example, P Jones, 1998; Lambert, 2000; Larkham, 2011). As Larkham and Lilley (2012) point out, this expert-driven culture of the 'great planner' has also become a dominant motif in British context and especially in the authoritarian regimes such as the USSR. Scott (1998) suggests that the new urban form that resulted from the 'project' of modernisation had to look regimented and rational. Le Corbusier's utopian blueprint for the future city (his *Plan Voisin* for central Paris) is often cited as being the embodiment of high-modernist urbanism, with its influence on the design of the state capitals such as Brasilia, Ciudad Guayana and Chandigarh (Holsten, 1989). Scott (1998, page 342) contends that in each of these cities, 'the order and certainty that had once seemed the function of God was replaced by a similar faith in a progress *vouchsafed* by (not only planners) but by scientists, architects and engineers'. As Matless (1998, page 209) points out, evoking the image of the phoenix 'enables a resolution of the tensions between history and modern rebuilding . . . a universal mythic creature rising from its own particular ashes connected local history and splendid newness'. Matless (1998, page 210) draws attention the cover of the architect Ralph Tubbs's (1942) '*Living in Cities*' and suggests that it would be a mistake if planners failed to take up the opportunity to plan for a cleaner, safer, less congested, more 'set-squared lined' and ordered future free from the nineteenth-century chaos and wartime bombing. It is interesting to note, in comparison, the many 'set-square lined' serried ranks of terraced houses in the late-nineteenth century bye-law developments.

As early as 1941, while the bombs were still falling, J.B. Priestley outlined in a celebrated special issue (complete with six naked, presumably impoverished, small children on the cover) of the *Picture Post* (1941), 'A Plan for Britain'. The magazine recalled the sudden end of the war in 1918: 'The plan was not there. We got no new Britain. . . . This time we can be better prepared. But we can only be prepared if we think now . . . the new Britain must be planned offered an initial blueprint for a fairer, happier, more beautiful Britain than our own'. Alongside the work of Donald Gibson and his plans for the rebuilding of Coventry and Thomas Sharp's plan for Exeter (Tait and While, 2009), these plans have been interpreted by Gold and Ward (1994) as being created by men of vision, charged with making reconstruction exciting, informative, progressive and idealistic. Commenting on Professor Abercrombie's role in Jill Craigie's film, *The Way We Live* (1946) – a film made to document Plymouth's wartime destruction and Abercrombie's vision for the post-war city – Tewdwr-Jones (2013) notes that the film tends to re-affirm the public view of the professional expert, particularly when the film's narrator states: 'no-one knew what the professor was up to'.

The increase in motor traffic in the 1930s in all British cities had caused both chaos and the creation of dedicated spaces for traffic, pedestrians and civic parks and squares that formed the basis of many new plans (Gold and Ward, 1994). This began before the onset of war, of course. However, armed with sophisticated statistical analyses of traffic censuses, Alker Tripp's (1938, 1942) theories and Abercrombie's approach to the replanning of the Bloomsbury area of London, for which he proposed a traffic-free precinct based on the Inns of Court (Forshaw and Abercrombie, 1943, pages 50–53), the traffic engineer could plan the roads of the future with unchallengeable certainty. There were increasing national anxieties in the post-war period over the economic and social impact of poor roads, particularly with inner-city motorways whose main purpose was to increase the speed of commercial and industry traffic. For example, the government-sponsored Road Research Laboratory estimated the costs of congestion to be in the region of £250 million in 1959, five times higher than the annual roads budget (NA CAB / 129 / 1010, 25 Mar, 1960).[6] This reflected the rising influence of highway engineers, culminating in the Buchanan Report of the early 1960s and the fashions for traffic-free precincts and ring roads.

There was some spatial and thematic differentiation in the portrayal of these new urban visions. For example, plans often sought to address important issues such as housing and the transport infrastructure that was required to facilitate the building of a 'Better Britain'; however, many also included new civic centres and similar facilities, for which local populations saw less immediate need (Larkham, 2004). Even where there was less physical destruction because of bomb damage, the plans also appear to have reflected a burgeoning concern for conservation and preservation as well as redevelopment (e.g. Pendlebury, 2003). There were certainly strong conservationist elements in Thomas Sharp's plans for the reconstruction of the heavily bombed damaged Exeter (see, for example, Pendlebury, 2003, 2009) and the less damaged Durham, Oxford, Salisbury and Taunton (Larkham and Lilley, 2003; Pendlebury, 2003, 2004) and Chichester (Larkham, 2009). Such a perspective was typified by Sharp's clarion call that 'the watchword for the future should be – not restoration but renewal', but this ran counter (in some respects) to the advice permeating the 1947 government *Advisory Handbook on the Redevelopment of Central Areas* (MTCP, 1947) promising a more 'technocentric' and progressive approach to replanning.

Elsewhere, in his *Oxford Replanned,* Thomas Sharp (1948, page 16) argued that 'planning . . . is not simply machinery for incubating brave new worlds; it is a technique which provides that new developments . . . shall not disrupt ancient practices or traditional forms where those are good'.[7] Such conservative approaches were evident in other parts of Britain. In Portsmouth, for example, a redevelopment plan of February 1946 sought to retain the partially destroyed Guildhall with a view to making it the city's new social and cultural focal point (see Larkham, 2014a). Hollow (2012, page 580) argues that, despite their utopian and overtly optimistic tone, these ideas could also be interpreted as an attempt to make a 'direct impact on people's lives by providing positive visions' of a possible post-war future.

These perspectives were not exclusive to Britain. Reflecting Ward's (1999) argument that planning ideas can 'travel' between importing and exporting nations, Chapman (2005, page 248) discusses how Austen Harrison and Pearce Hubbard, the two Liverpool-trained consultants appointed by the British Colonial Office, did much to create 'good new civic spaces into the tight-knit street patterns of Valletta' (on this general issue, see also Nasr and Voilait, 2003; Cook *et al.,* 2013; Ward, 2017). Although this perspective found expression in Coventry, where Donald Gibson, the City Architect, and his department used the ideas of Le Corbusier's (1929) *City of Tomorrow* (Johnson-Marshall, 1966), here as in most places a mixture of local politics, restrictions on labour and shortages of capital and delays in finance (Tiratsoo *et al.,* 2002; Flinn, 2012, 2013) watered down the practical translation of this visionary approach. Despite these limitations, radical proposals continued to emerge after the war.

One notable extreme expression of post-war modernist proposals was Geoffrey Jellicoe's (unrealised) plan in 1959 for a 'Motopia' New Town sponsored by the Glass Age Development committee of the Pilkington Glass Company. This town of 30,000 inhabitants would be built around geometric grids of elevated roads rising 50 feet into the air with roundabouts at each intersection and the ground level reserved for public parks and canals and the town's residents could travel to and from work by water bus and by airport-style travellators (Moran, 2010). A less extreme version in terms of architectural vision, but no less radical in terms of built form affecting way of life, was the idea of small-scale communities of dwellings arranged around miniature 'village greens' with communal facilities including kitchens, promulgated by Professor Sir Charles Reilly. Introduced in his plan for Birkenhead and adopted in only a small scale in the Midlands (Larkham, 2006; Harwood, 2008) this vision also had little direct impact.

Furthermore, once the lustre of the visionary plans/exhibitions and the national mood of euphoria waned, everyday realities of public disengagement and apathy rose to the surface (Hasegawa, 1999). While the 'boosterist' and place-promotion rhetoric of some wartime and post-war planners sought to foster a political climate for obtaining public consensus for the proposed newly reconstructed urban centres, Cowan (2013) argues that public interest tended to dissolve quickly after the end of the war. Rather than the 'Brave New World' of the 1946 'Britain can make it' design exhibition and London's 1951 'Festival of Britain'-inspired South Bank of the River Thames, most people were perhaps more concerned with a supply of good, traditional housing, brimming with all the modern conveniences (Atkinson, 2012; Hollow, 2012).

Communicating ideas

Work continues to explore how new ideas of imagining a new urban future were instilled in the reconstruction plans and how these plans were received. A focus on the reception – or consumption – of some of these ideas is a significant (yet underdeveloped) facet of the broader debate surrounding the reconstruction of war-torn British cities. In recent years, more consideration has focussed on the 'everyday geographies' of those subject to the plans, with a small (but growing) body of research exploring the exhibition of some of the plans and their relationship between propaganda and consultation (e.g. Lilley, 2004; Larkham and Lilley, 2012). Thomas Sharp's presidential address to the Town Planning Institute in 1945 is instructive of the prevailing attitude: he did not deny that it was incumbent on planners – as 'experts' – to explain to people what was being planned for them, including the 'right to comment on plans, to require alterations to them and, if necessary, to reject them'. What he rejected, however, was the right for people to participate in the 'act of planning', in other words before the draft plans had been drawn up, the notion that the planner should be putting the public's wishes into technical form (Sharp, 1946). Abercrombie held similar views:

> In the first instance a plan is prepared independently and almost in secret by the technician. He is given complete freedom to prepare a plan on whatever lines he thinks fit, having access to whomsoever he thinks it necessary to consult. He does not submit that plan, at any stage, to any local authority or Government department for their scrutiny or approval.
>
> (Abercrombie, 1949, page 10)

Although there was a distinct and noticeable move towards public communication and education embedded within the wartime reconstruction plans (and their accompanying literature), arguably less consideration centres on precisely *how* members of the public could shape these future cities (Larkham and Lilley, 2012). Exhibitions, books, broadcasts and a whole raft of promotional material disseminated by professional journal articles conveyed a broadly consistent and positive message about the post-war future. Many of these exhibitions were very tightly focussed and place-specific. Attention focussed on ameliorating the worst excesses of the pre-war city and the impacts that the wartime bombing had inflicted on certain places. Others were more generic, and some travelled around the country; notable examples of this type of approach included '*Living in Cities*', designed by the architect Ralph Tubbs for the British Institute of Adult Education (Tubbs, 1942); '*Homes to Live In*', '*The Englishman Builds*' and '*Plan Scotland Now*' (Larkham and Lilley, 2012). It could be reasonably argued that town planning exhibitions, films and promotional publications of the early to mid-1940s were entangled within a broader (albeit, perhaps unfocused) culture where displays of wartime 'propaganda' were an accepted facet of everyday life (Tewdwr-Jones, 2013). The reconstruction plans, maps and exhibitions were imbued with a sense of civic pride and boosterism, with towns and cities jockeying for position in a time of economic and social uncertainty (Shapeley, 2012). Drawing on the example of Worcester, Larkham (1997) suggests that the importance of these types of sanitised images revolved around two distinct functions: first, they were enmeshed in the processes of civic boosterism and place marketing; and second, they were significant in terms of how they sought to control and order aspects of residents' everyday lives. Few previous planning documents had so clearly sought to do this.

The '*Manchester of the Future*', a 'Comprehensive Exhibition of the City's Redevelopment Plan' based on the ideas of the City Engineer, Rowland Nicholas (Nicholas, 1945, page 18), promised a city of more light, air, open space and widened roads. The Minister of Town and Country Planning opened the exhibition in July 1945, and within six weeks, 125,000 people had attended (Perkins and Dodge, 2012). Shapeley (2012) explains that even as late as October 1965, Manchester Planning department held a three-week long exhibition (attended by 10,000 visitors – including the then Prime Minister, Harold Wilson) to 'showcase' the city's reconstruction ambitions. As with elsewhere, this involved producing visions of an optimistic post-war future, with white models portraying a bright, shiny future and a desire to eviscerate the dirty, defiled, cluttered and unhygienic urban core that had characterised the nineteenth-century city.

Shapeley (2012) suggests that public exhibitions were also displays of municipal 'boosterism'. In addition, however, this culture of 'consultation' through exhibitions was precariously balanced between controlling certain elements of human experience – essentially, how people viewed and interpreted the plans/models – and allowing visitors to experience, digest and re-interpret these representations for themselves. Overall, though, one might suggest that visitors to the exhibitions were thus consumers of types of prescriptive choreographies embedded within the style and layout of how the exhibitions displayed information and guided the visitor around (see Hornsey, 2008, 2010).

The realities of reconstruction

There is an increasing acceptance that the aftermath of war encouraged what was already a powerful discourse for political, social and economic change and for a more rational and scientific move towards the reconstruction of British cities (Hebbert, 1983). Barnett (1986) has pointedly argued that Britain during and after the war had made a profoundly mistaken choice in being seduced by the elite-driven rhetoric of 'New Jerusalemism' (Stevenson, 1991) and did not give due consideration of economic reconstruction as an unsentimental priority over social reconstruction and the creation of a welfare state. In this sense, the rather restrictive nature of state-sponsored planning largely curtailed the influence of individual planners was limited because they were part of much broader and more complex political and economic webs.

One of the immediate problems facing blitzed cities was the shortage of building materials, which remained rationed until the early 1950s. Structural steel was a problem, as was acquiring building licences (as a way of securing structural steel). Steel was allocated to bombed cities by Ministerial committee and was exported even to Australia to gain much-needed foreign currency. Ministerial and legislative hurdles further compounded this issue. Although the 1944 and 1947 Town and Country Planning Acts presented local authorities with increased powers to initiate the process of rebuilding, both were vague in how they affected towns and cities (Flinn, 2012, 2013). The 1944 Act, a hurriedly introduced piece of Coalition government legislation, was designed with the principal purpose of providing details over levels of compensation for war-damaged property during and after the war. This was extended to slum clearance because of professional input (claimed specifically by Herbert Manzoni, Birmingham's City Surveyor and Engineer). Lewis Silkin (and colleagues) amended the Act, in his role as the post-war Labour Minister for Town and Country Planning. This clarified details regarding the methods of public land acquisition and for calculation of compensation (Ward, 2004).

Furthermore, despite demobilisation immediately after the end of the war, the nation's labour force was severely hit, and the nation's capacity to export was desperately lacking even by the end of the 1940s (Judt, 2013). This perilous state of the post-war economy had an effect of creating more pressure on imports and consumer spending with the US being the obvious source for many of the desirable consumer goods (Marwick, 1968). Unfortunately, for Britain, the introduction of Marshall Aid had rather limited impact on investments or the modernisation of industry because 97% of the financial aid was reserved for debt repayment and that Britain was also a debtor nation and dependent on foreign capital and imported raw material (Judt, 2013).

The particularly harsh winter of 1946–1947 further stymied the British export drive, while the 'convertibility crisis' further undermined any concerted attempts towards reconstruction, leading to several substantial investment cuts. In a more pointed missive against post-war economic stringency, Jewkes commented in *Ordeal by Planning* (1948) that tight fiscal controls, like those imposed on acquiring building licences, were a source of immense frustration to public authorities and communities alike wishing to proceed with the rebuilding (see also Larkham, 2006). Such a situation lends weight to Mandler's (1999) broad argument that planners and planning had a rather limited role in shaping change in the immediate post-war years, and that the reconstructed built environment was as much the consequence of market forces as of the strict implementation of the 'grand' ambitions of planners (and other professionals and decision-makers). As economic prosperity developed during the 1950s, however, the property sector regained its strength and there were profits to be made in coming forward with city centre redevelopment schemes to the extent that local authorities became more closely aligned with the private sector and the funds it could muster (Samaurez Smith, 2015). By the early 1960s, Ministerial guidelines, *Town Centres; Approach to Renewal*, suggested that local authorities should seize the opportunity to 'collaborate' more closely with private developers and that 'redevelopment is coming from private developers and the local authority has to move fast to keep pace with them' (NA HLG 136/20, 1962, page 6).

Recognising actors in the process of plan-making is a theme explored and developed by several authors. Essex and Brayshay's re-working of Latour's 'Actor-Network-Theory' has begun to unpick and outline the ways in which these 'actors' were involved in the decision-making processes that led to the replanning of certain towns and cities (Essex and Brayshay, 2007, 2008). There is also another (arguably underdeveloped) dimension to the 'messiness' of implementing some of the ideas for these numerous 'paper cities': the distance that existed between the concrete policies that emerged from these plans and the tensions over the scale and speed of implementation. Most notably, perhaps, local action embodied within a city plan was often hamstrung by the structures enshrined in local and regional planning documents (Essex and Brayshay, 2005); by national planning frameworks and directives (Malpass, 2003); and because of disputes between different government ministries and representatives of bomb-damaged cities, including setting priorities for finance and materiel (Flinn, 2012, 2013; Greenhalgh, 2017).

It is also worth emphasising that planning, as a profession, was young, and there were relatively few professionally qualified town planners, with much 'planning' activity undertaken by architect-planners or engineers. During the war years and until widespread demobilisation, many younger professionals were away on military service; quite a few did not return. In some cases, the status and prominence of the 'plan makers' working with local authorities ensured that there was no real need to seek advice from expensive and eminent consultants; some authorities appointed consultants more for reasons of local competition than need (Warwick versus

Leamington Spa, perhaps). Local interest groups such as civic societies, local individuals or other organisations produced a small number of plans. The Ministry commissioned most of the regional plans, and Abercrombie had a hand in many of these. Commercial consultants often vied for the profitable commissions from cities, and highway engineers, city surveyors and others competed for resources with architects and public health officials. Clough Williams-Ellis, owner-architect of Portmeirion (and historian for the Royal Tank Regiment), for example, was clear that this important task should fall squarely at the feet of architects (Larkham, 2014a). On the other hand, he proved unable to complete the plan he undertook for Bewdley (Larkham and Pendlebury, 2008).

In 1942, the Institute of Municipal and County Engineers also laid claim to the responsibility of planning by writing to local authorities, suggesting that 'qualified Engineers and Surveyors of Local Authorities are fully competent to continue and complete the preparation of planning schemes' (quoted in the minutes of Dover's Post War Development Sub-committee, 23 Nov, 1942: East Kent Archives Centre). As Fyfe (1996) discusses in relation to the replanning of Glasgow, two plans were put forward for city and a team of regional planners, commissioned by the Scottish Office and led by the eminent Patrick Abercrombie, ventured in 1944, culminating in publication of the Clyde Valley Regional Plan in 1949 (Abercrombie and Matthew, 1949). Four years earlier, however, Glasgow Corporation had published its own plans, drawn up by the City Engineer, Robert Bruce, which proposed a thorough examination of the need for fundamental improvements in the transport, housing and land-use requirements of the city over the next fifty years (Bruce, 1945). In addition, reviews of reconstruction plans came from 'planning's' allied professions, including the *Architects' Journal, The Builder, Architectural Review* and *Architect and Building News*. This spread of journals also gives some indication of the early development of the planning profession, reflecting the fact that many plan authors were not town planners by training.

There also appeared to be some debate *within* the Planning Ministry over the question of which profession would be the most suitable for task of rebuilding, with Ministry staff commenting acerbically and very personally on the abilities of surveyors and engineers to plan properly (Hill to Vincent, 14 Jun, 1941: NA, HLG 71 / 760). The Ministry of Transport (or Ministry of War Transport) had overall responsibility to approve all roadway designs, alongside the requirements of other ministries to deal with statutory undertakings. However, local authorities in the blitzed cities had to agree on a plan with the planning ministry. But grand ambitions could be delayed further if there was disagreement from the Ministry of Transport over the specificities of road layouts, or the Ministry of Health over infrastructure or zoning, as well as from other government ministries such as Works which might be responsible for civic building (Larkham, 2014a). While the Planning Ministry's technical planning staff had a remit for making decisions nationally, the plans made by local authorities also hit significant obstacles and delays in the rebuilding ensued (Tiratsoo, 1990). Furthermore, plans and exhibitions organised by several bombed (and unbombed) towns and cities raised general levels of public expectation (Larkham and Lilley, 2012), which could not be fulfilled for a whole host of interrelated reasons.

Living with reconstruction

Buildings depicted in many of the outline development plans were bland boxes, and often criticised as such (for example, Thomas Sharp's biographer Stansfield (1981, page 156) called the views of buildings in his replanned Exeter 'insipid boxes which had no townscape qualities'). Yet many plans explicitly noted the position and scale of the future buildings and hinted at later design treatments. Following the enactment of the seminal 1947 Town and Country Planning Act and the requirement that local authorities produce a more technocentric form of Development Plan' ensured that the (sometimes) lavishly produced and highly visual reconstruction plans so widely produced in the early years of the decade tended to diminish. Hence, conveying what 'planning' meant to a less-interested public became more difficult

Acknowledging the shortage of labour, materials and finance, it would also be erroneous to assume that strong local opinion existed in favour of the planners' ideas. In Bristol, for example, a poll organised in early 1947 by the Local Retailers' Federation discovered that around 400 people were in favour of the new Broadmead shopping centre, whereas the calls from a staggering 13,000 people who wanted to see the old shopping centre rebuilt were, to some extent, ignored (Hasegawa, 1992; Larkham, 2003). As with other badly damaged cities, including Exeter and Plymouth, the outcome of building was modernist-inspired with a proliferation of rectilinear precincts, ring roads, social housing estates and re-engineered town centres (Kynaston, 2007).

Considerable attention has centred on the narrative of redevelopment of British cities following the end of the Second World War. Much of this has been captured in official discourse (plans, maps, council publications, models and so on) that often describes the city through the 'planners' gaze' (Haffner, 2013). For example, several accounts detail how the images contained in many plans and associated documents influenced key decision-makers and their visions of post-war replanning. In this analysis, improvements to the urban way of living represent treatments and cures for the city as an 'ailing body'. In other studies, Matless (1998) has described how planners promoted certain codes of moral conduct involving the production of images and representations that had the intention of making city appear legible to its occupants, stressing the desirability of new ways of urban life. Elsewhere, Hornsey (2010) explores how ideas of pedestrian management, control and order were entwined in the mid-twentieth century plans for London and how these notions continue to impact on pedestrian movement in other British cities. The official documentation demonstrates how different to post-war 'new Britain' was intended to be – even if the 'new Jerusalem' did not come about. But this perspective is only partial: key actors, principally residents, are largely marginalised from this story.

Summary

Following the end of the Second World War, Britain needed to build many new homes and rebuild blitzed city centres. Unbombed towns and cities perhaps had less *need* to rebuild, but *wanted* to do so. More British towns and cities were replanned in a shorter period than ever before, and this involved the widespread introduction of radical new concepts of architectural and built form, and a technocratic, scientific and data-led approach to planning became dominant. While much has been written regarding the form, product and process of reconstruction planning, this has focussed largely on 'official' municipal, governmental and professional perspectives. Significantly, less attention has been paid to the *experiences* of the people inhabiting or using the then newly configured urban spaces. Although many authors are beginning to be attuned to the different ways humans encounter planned spaces, relatively few studies (to date) have examined this empirically in relation to the 'Modern' built environment. Instead of reading these plans and images as neutral representations based on rational, technical processes of surveying and mapping, consideration should be focussed on how these spatial concepts connected and interacted with everyday urban life. Chapter Three moves on to discuss the different methodological approaches used to capture everyday life in the post-war city.

Notes

1 Although, as Larkham (2013) notes, there may be some discrepancy here and there should be a 15–30% margin of error (including double counting of separate damage incidents to same property).
2 Between 1924 and 1938 the number of fatal road accidents reported in England and Wales had increased by 77% (from 3,269 in 1924 to 5,809 in 1938). Non-fatal (so far as report) road accidents increased by a staggering 96% (see Tripp, 1938).
3 It should, of course, be noted that the Churchill administration did initiate some forms of social reform, perhaps most notably the 1944 Education Act.
4 The subsequent creation of the similarly short-lived British Ministry for Town and Country Planning and its successors played a major role in shaping a planned response to wartime destruction and the changing priorities of a changing society; and as Larkham (2011) acknowledges, its 'Planning Technique' section also established new approaches to data-driven planning.
5 John Reith had set up the British Broadcasting Corporation during the 1920s and 1930s (see Reith, 1949). He was created Lord Reith in October 1940.
6 The National Road Research Laboratory (est. *c.* 1933, later the Transportation and Road Research Laboratory) was charged with the task of responding to issues

of ensuring effective urban traffic management, especially in its development of a formula for 'the traffic capacity of weaving sections of roundabouts' (cited in Redknap, 2004, page 38).

7 Though Sharp's 'Building Britain' poem, written in 1941, does much to convey a sense of progressive thinking:

> 'There is so much to plan for, to prepare for,
> A whole shining world is possible . . .
> Where in brightly coloured lanes
> Along ordered avenues
> The traffic towards the future will flow easily, smoothly'.
> (Sharp, 1952)

3 Disaster and opportunity
Replanning Coventry and Birmingham

Introduction

Drawing on 'official' published accounts, including national and local archival material, council minutes and newspaper reports, this chapter explores in some detail the pre-war conditions that existed in Coventry and Birmingham and the planning response to the wartime situation. While these two cities are not 'typical' of the UK's response, Coventry became an international icon of destruction, reconstruction and reconciliation. Birmingham became equally well-known, perhaps infamous, for its traffic-dominated response to post-war reconstruction. Detailed sources exist for both cities. Coventry has been well-studied, but the dominance of its plan-led approach merits review for our perspective, while Birmingham's approach to post-war renewal has been studied in the recent past and, unusually eschewing an overall reconstruction plan, merits comparison. Following a brief description of the levels and extent of the cities' wartime bomb damage, the chapter moves on to describe the professional background and influence of the dominant figures Donald Gibson and Herbert Manzoni in shaping post-war planning ideas for the two cities, and it closes by outlining how these ideas were communicated to members of the public.

Pre-war context

Birmingham was not a large medieval town – it is primarily an industrial creation. By the late-nineteenth century, however, Birmingham assumed a national and international reputation as a prototype of successful municipal governance, and the city's administrative and physical development during the Victorian and Edwardian eras have been well-covered (see, for example, Briggs, 1952; Sutcliffe and Smith, 1974; Hopkins, 1989 (revised 1998); Skipp, 1983; Cherry, 1994; Foster, 2005). Such ambition was effectively marked out in the 1870s when the then Lord Mayor and MP for Birmingham between 1876 and 1914, Joseph Chamberlain, expressed his 'public-spirited capitalism' (Mah, 2012, page 8). This was described by some commentators as invoking the spirit of the 'civic gospel' (Briggs, 1952) or the 'municipal gospel' (Hennock, 1973) to guide him in his quest to improve the city's physical infrastructure. Such an approach sought to remove the deplorable 'slum'-type housing conditions, death and disease that had proliferated during the nineteenth century industrial expansion of the city (Tiptaft, 1947).

Such reforming zeal found political and practical expression through Chamberlain's insistence on the development of public services through the municipalisation of the city's gas, water and sewerage systems. New libraries, museums, parks and galleries were also created (see, for example, Chinn, 1999), and the City Improvement Scheme – made possible through the enactment of the Artisans' Dwellings Act of 1875 – involved considerable slum clearances of insanitary homes and the creation of new streets, shops and services (Upton, 1993), including Corporation Street, a new 'breakthrough street' designed to rival Parisian boulevards. Joseph Chamberlain's Corporation Street scheme was a 'very early example of modern town planning, in that the plans included the demolition of a large, obsolete area, the re-housing of the people who had lived there and the creation of a broad new shopping street and commercial centre' (Manzoni, 1968, page 1). It could also be characterised as representative of planning projects in the length of time taken to complete, and the lack of re-housing provided for displaced residents (Yelling, 1986).

However, notwithstanding the optimistic reputation of Chamberlain's civic aspirations and the successful delivery of some aspects, Birmingham City Council largely ignored the need to replace most of the housing it had demolished, and housing remained the most conspicuous oversight in official municipal policy until after the First World War (Mah, 2012). Slum clearance was therefore a central concern for the City Council during the 1930s, to the extent that Manzoni later claimed that 'the idea of redevelopment in Birmingham was born in 1936' (Manzoni, 1955, page 90). Birmingham did make progress in other areas of municipal planning policy. The first Town Planning Schemes approved – in 1913 – under the Housing, Town Planning Act of 1909 were in Birmingham, and the city had prepared several schemes for suburban areas by the onset of the Second World War. It was involved in the Midland Joint Town Planning Advisory Council, covering the counties of Staffordshire, Worcestershire and Warwickshire (Cherry, 1994). The first ideas of an inner ring road can be traced back to a visit by the then City Engineer and Councillors to Germany and Austria in 1910 (Borg, 1973, page 51).[1] Proposals for an inner ring road, if not actually formally proposed, were mooted as early as 1917, designed to join the ends of the wide radial routes that were then planned (Manzoni, 1968, page 2). Other road improvements had also been planned but halted by the war, including a dual carriageway through Deritend and Digbeth (south-east of the city core; permitted in 1935).

Municipal administration in a rapidly growing city was itself growing rapidly. A civic centre had been proposed informally during the First World War, and an architectural competition held in 1926 for a site immediately to the west of the city core, on Broad Street (Foster, 2005, page 144). The Birmingham Corporation Act (1936) included powers for the compulsory purchase of land for the civic centre and, owing to the war and post-war

CORPORATION STREET, BIRMINGHAM.

Figure 3.1 Birmingham in the 1930s
Source: Reproduced with permission of John Lerwill

financial rigidity; requests were made to extend these powers into the early 1950s.[2] Part of one office building, Baskerville House, was completed in 1940 in the Classical style and white Portland limestone selected for the civic centre (Sutcliffe and Smith, 1974, page 449). Despite these large-scale planning interventions, Birmingham's pre-war built environment was characterised by street layouts inherited largely from the industrial age and by many manufacturing and engineering buildings (Cherry, 1994).

In 1948, the West Midlands Planning Group[3] – a group that brought together a number of influential people, acting unofficially and largely in a voluntary capacity, to consider the post-war reconstruction in the West Midlands – produced a seminal report entitled *Conurbation: A Planning Survey of Birmingham and the Black Country* (West Midland Group on Post-War Reconstruction, 1948). Here it was reported that the 'population of the West Midlands Conurbation was increasing over the first 40 years of this century at a rate far higher than occurred over the country as a whole' (page 85). High fertility rates rather than in-migration were identified as being the main reason for this increase, and there were justifiable concerns about how to accommodate increases in population into the existing built fabric of the West Midlands. In other publications, the pre-war conditions of Birmingham were succinctly described in unfavourable terms (Figure 3.1):

The central fringes of most of our older cities more so with an industrial background, are usually ill-planned, over-crowded and insanitary.

The street pattern is out of date and unfit . . . there is a shortage of open space, the land is used in a most uneconomic fashion . . . industry is intermingled with living accommodation to such an extent that one can't tell where one begins and the other one ends.

(*Birmingham Evening Mail*, 1959, page unknown)

The pre-war position in Coventry

Unlike the mid-late-nineteenth century expansion of Birmingham, Coventry's growth was rather more sluggish. The ancient Lammas and Michaelmas grazing land encircling the Western edges of the city centre was not made available for building until the late-nineteenth century. This, together with the location of the fourteenth-century city wall, inhibited concerted outward expansion of the city for some centuries (Fox, 1947; Stephens, 1969). Between 1861 and 1871, the city lost population and international competition adversely affected the city's traditional watch-making and ribbon-weaving industries (*Architectural Design*, 1958, page 474). Following the development of sewing machines, bicycle manufacture and the motor industry, however, the city experienced a burst of industrial growth during the first half of the twentieth century and there were major manufacturing sites deep within the city core as well as on the urban fringe (Gregory, 1973; Newbold, 1982).

In planning terms, and like Birmingham, there were some notable developments. In 1925, Ernest Ford, the then City Engineer, proposed a series of inner ring roads to siphon vehicular traffic away from the city centre, and a six-mile-long southern outer by-pass (A45) intended to take London to Birmingham traffic was completed by 1939 (Gregory, 1973). There were also selective attempts at slum clearance of the congested central area still dominated by decayed timber-framed buildings, which had made possible the pre-war attempts to widen Corporation Street (1931), thus creating 'active' shop frontages, and Trinity Street (1937), resulting in new thoroughfares carving routes through the seemingly out-dated urban core. Both new streets also allowed the construction of large department stores, such as the Cooperative Retail Society in West Orchard (1931) and for Owen Owen (1939) on the west of Trinity Street (Figure 3.2).

Following the election victory of the Coventry Labour Party in November 1937, the City Council faced some serious problems. The rapid early-twentieth century development of the engineering, electrical and automotive industries, and their accompanying attractive wage rates caused high net in-migration into the city (Gregory, 1973, page 83). This surge of in-migration of a young workforce, together with a relatively high birth rate, meant that the city's population of nearly 70,000 in 1901 had grown to over 204,000 in 1937 (Richardson, 1972). This early-twentieth century expansion meant that industry was undergoing 'constant change, (with) one craft displacing another as new inventions were taken up' . . . leading to a dominance of

Figure 3.2 Coventry in the 1930s
Source: Reproduced with permission of Rob Orland

people employed in the 'motor, aircraft and machine tools industries' (Coventry City Council, 1958, page 6).

The city had a rather narrow economic base and a deficiency of 'black coated [office] workers' (Coventry City Council, 1958, page 6); this left the city with a below-average proportion of 'professional, technical and administrative classes' (Gregory, 1973, page 83). For some commentators, the pace and scale of Coventry's growth ensured that the 'city outstripped its physical equipment and social development' (*Architectural Design*, 1958, page 474). In a sense, the city had a 'backlog of housing, services and facilities issues needing resolution' (Fischer and Larkham, 2018, page 83). In the city centre, there was a patchwork of small freehold interests, and commercial pressures increased on central sites during the early part of the twentieth century. This had resulted in piecemeal, speculative growth and high land values making compulsory purchase of sites – especially in the absence of national statutory planning powers and, crucially, funding, for comprehensive redevelopment – increasingly difficult during the 1930s (Gregory, 1973).

Before the onset of the Second World War, therefore, Coventry city centre was made up of a high-density assemblage of industrial workshops and factories interspersed with business uses, offices and shops, all overlain on a largely medieval street pattern. There were significant housing shortages and appalling housing conditions, and the city centre was deficient in shopping and office accommodation (Redknap, 2004). The area around Broadgate, 'the central place' within the city centre, became an increasingly congested site of activity before the Second World War with 'buses, trams, commercial and car traffic' causing 'nuisance and danger' for pedestrian shoppers (Gregory, 1973, page 86). The city's 'three spires' (St Michael's Cathedral, Christ Church and Holy Trinity Church) had dominated the city skyline since the fourteenth century and remained as historic symbols (Stephens, 1969). However, the clear pressures to redevelop, remove and remodel aspects of Coventry's industrial and medieval past were recognised before the start of the Second World War (Gregory, 1973).

While it is invidious to consider the reconstruction in Birmingham and Coventry as wholly the product of the senior staff involved – Herbert Manzoni and Donald Gibson, they rightly have been considered very influential, and they – and several key individuals with whom they interacted – are introduced here.

Birmingham: enter Herbert Manzoni

Born in Birkenhead, Herbert John Baptista Manzoni (1899–1972) was educated at the Birkenhead Institute before moving on to Liverpool University. After serving in the First World War (1916–1918), he relocated to Birmingham in 1923 where he was appointed Engineering Assistant in the Sewers and Rivers Department. He became the Chief Engineer in the Sewers and Rivers Department, then Deputy Chief City Engineer and Surveyor; and at the age of thirty-six, he became the City Engineer and Surveyor in 1935. He held this post until his retirement in 1963. Manzoni was awarded the CBE in 1941 following his efforts to marshal an effective air raid precaution programme for Birmingham; he was knighted in 1954 (Caulcott, 2004).

Manzoni did have some architectural training, though he never fully joined the Royal Institute of British Architects (RIBA) (Sutcliffe, 1967–1969). He was, however, active in delivering talks on post-war reconstruction and housing to local interest groups in the West Midlands (e.g. the Rowley Regis Rotary Club); the Birmingham and Five Counties Architectural Association and to national organisations on the issue of mass-produced housing (at the Association of Municipal Corporations, London (see Larkham, 2007)) and on post-war planning and reconstruction (the Institution of Civil Engineers: Manzoni, 1942). Lewis (2013, page 35) provides an informative account of how Manzoni was also invited to advise Sheffield on various reconstruction plan options; he was 'regarded as a leading member of the Institute of Planning and the Institute of Civil Engineers'. Manzoni was a member of the Town Planning Institute and member of the Town Planning Committee of the Association of Municipal Corporations; also President of the Institution of Municipal Engineers (1943–1944), the Institution of Civil Engineers (1960–1961) and the British Standards Institution (1956–1958) (Caulcott, 2004, page 577). Interestingly the Institution of Civil Engineers juxtaposed his portrait with that of Sir Winston Churchill on the front cover of the book accompanying its exhibition 'Visions of reconstruction 1940–1948' (Baldwin, 2005).

This was a time when 'a great chief officer could exercise enormous personal power and influence. Manzoni came into this category': 'elected members relied on [his] advice', and his use of 'far-sighted and controlled judgement' earned him 'legendary respect' amongst his peers (Caulcott, 2004, page 577). Sutcliffe (1975, page 20) offers a cooler interpretation of Manzoni's more dogmatic character traits: when interviewed by Sutcliffe (1967–1969), it that Manzoni 'preferred to give a lengthy exposition without interruptions from the questioner'. Ward (1989, page 24), however, provides a more balanced perspective, believing that Manzoni was a more urbane character and 'not an ignorant technocrat . . . he was a dedicated public servant to his city and used his best wisdom . . . to solve its traffic problems'. Nevertheless, as a person, he remains a rather shadowy character, and his department was structured and run in a very traditional manner.

Manzoni was forceful in his belief that engineers were suited to lead 'town and country planning' because of their greatest accumulation of experience in the subject and the 'right sort of training' (Manzoni, 1942). However, there has been criticism of Manzoni's own approach to town planning. He drew withering derision from local architect, John Madin, who suggested that Manzoni's understanding of effective town planning was inherently limited because he was 'a road engineer' (Sutcliffe, 1967–1969). Interpretations of Manzoni's influence on Birmingham are rather mixed. Some have interpreted Manzoni's 'scorched earth' policies of 'razing whole districts to the ground' (Steadman, 1968, page 1383) together with his role and achievements in a very critical fashion:

> The engineer's eye of Manzoni saw his job as complete once roads had been driven over freshly cleared ground. The new opportunities for development created by each road project were dealt with piecemeal, resulting in a lack of aesthetic consistency. Paradoxically, the City Centre suffered from a lack rather than surfeit of planning for the buildings themselves. Getting things built rather than ensuring their aesthetic value became the priority.
>
> (Parker and Long, 2004, pages 7–8)

Chinn (2013) offers a similarly damning perspective of Manzoni's utopian approach to replanning: 'as a powerful and inspirational leader, Manzoni cared deeply for the city – but like so many professionals, he did not understand nor seek to understand either working-class culture or the need for proper consultation with the citizens of Birmingham'. It seems, therefore, at least on this evidence, that Manzoni preferred clean, modern designs and had little time for older architecture and urban layouts (Larkham, 2016). Manzoni sympathised with earlier efforts to instil a 'civic gospel' approach to the governance of 'reconstruction'. Yet Manzoni's influence was mediated through other actors and processes, especially given the influence of the other local authority committees and particularly Frank Price, the chairman of the powerful Public Works Committee from 1953–1959 (see Price, 2002).

Frank Price (see Price, 2002) was a local man, who stood for election to the Council in 1949 aged only 27. Five years later, he was chair of the Public Works Committee, thus wielding enormous influence over the city's reconstruction (and other development-related issues). It is said that 'most of what Birmingham is now trying to get rid of is due to him' (Coulson, 2003, page 127). Arguably, more than any other local politician, he expressed an ambition for the city to be the leader in urban renewal, so, as he frequently reiterated, 'we can truthfully say [Birmingham is] leading our country if not the world' towards post-war reconstruction (Price, 1960, page 4). He once described the city as having an 'air of stagnation' (Price, 1960, page 4); and that this condition could be changed by the agency of an effective planning programme.

He has been described as 'outspoken . . . an aggressive, controversial local politician' (Marriott, 1967, page 125). Describing himself, he said 'as a committee chairman, I achieved quite a lot by histrionics, flamboyance and what I call gut fighting' (interviewed in Sutcliffe, 1967–1969). In 1959, however, and apparently as a result of his high-profile activities promoting Birmingham's redevelopment and activities with property developers, he was offered a job, later a directorship, with a major property developer, Murrayfield Real Estate and stepped down as chairman of the Public Works Committee. 'To be a director of Murrayfield was a far more interesting and better paid job for Price. . . . Many of his Labour colleagues thought that he had sold the pass when he joined Murrayfield' (Marriott, 1967, page 125). However, he became Lord Mayor in 1964–1965 at age 42 and oversaw the opening of the Bull Ring by the Duke of Edinburgh – who apparently referred to Price as 'here comes trouble'. He was Leader of the Birmingham Labour Party. He was knighted the following year, aged 43, before becoming Managing Director of Murrayfield in early 1967. He left the Council in 1974.

Coventry and the appointment and influence of Donald Gibson

In late-1930s Coventry, responsibility for city planning lay firmly within the remit of the City Engineer, Ernest Ford (see Johnson-Marshall, 1966). There were early signs of a shift in power when the new Labour-controlled Council took the decision to appoint an architect to improve the city. In short, the City Council required someone who shared its vision, ethos, aspirations, enthusiasm and commitment (Walford, 2009, page 134), or a 'young radical', as Tiratsoo (1990) suggests. 'Men of long service' were prohibited from applying for the position of City Architect; the age limit for the post was set at 40 (Coventry City Council Archives [hereafter CCA] Policy Advisory

Committee Minutes, 11 July, 1938). Donald Gibson became City Architect for Coventry in January 1939. It is widely acknowledged that he had an important personal influence on the city's later post-war reconfiguration. He also structured his new department in a radical manner, working as a group, giving individual credit where appropriate, with a more equitable remuneration scheme than was usual in municipal offices and recruiting 'interesting people' (Gibson, interviewed by Andrew Saint, NA, Architects' Lives C447/11, 1984). In a speech to the Royal Society of Arts in 1940, Gibson called for the creation of a Bauhaus-inspired institution to lead the way. This involved 'a combination of technical school, architectural school and town-planning department . . . the most gifted eventually becoming planners of the future environment' (Gibson, 1940a, page 1583). This is virtually what he had just created in Coventry.

Gibson graduated from the Manchester School of Architecture in 1931; he went on to work at the Building Research Station in Watford and taught construction at the respected Liverpool School of Architecture. He also studied under Professor Patrick Abercrombie. Several accounts detail the broad contours of Gibson's major work. There are, for example, several authoritative narratives, which explore his time as Deputy County Architect at the Isle of Ely (1937), where he explored the impacts of day-lighting and the design of classrooms and his move to Coventry (1938–1954) where he became the influential City Architect. Studies also detail his move to Nottingham in 1955 where he and his team pioneered the 'CLASP' flexible steel frame building method. He later moved to the War Office as Director-General of Works in 1958, and then he took a role in the Ministry of Public Building and Works in 1962, before his election as President of the RIBA (1964–1965) (see, for example, Campbell, 2004, 2006; Walford, 2009). Gibson was knighted in 1962.

At the end of February 1939, Coventry City Council's Policy Advisory Committee requested Gibson to devise plans for a new civic centre (*Midland Daily Telegraph*, 25 Feb, 1939). There was a desire to create a new civic area in Coventry before the war. It is clear from these early proposals that an overarching planning framework was required for the city: this should prevent piecemeal rebuilding from thwarting the future reconstruction of the city centre. Throughout 1940, Gibson, and his assistant and former pupil from Liverpool, Percy Johnson-Marshall, together with other members of the department continued to work on the plans for the reconstruction of the city, working largely in their own time (Gold, 1997).

The resulting Coventry of Tomorrow exhibition of May 1940 (Walford, 2009, section 4.4) paid explicit reference to Le Corbusier's (1929) contentious *City of Tomorrow*; in so doing, such an approach presented a clear indication that European Modernism was becoming an important influence. To generate levels of local public enthusiasm for these ideas, the department staged a publicity campaign in the city, concluding in the 'Coventry of Tomorrow' exhibition (*Coventry Standard*, 18 May, 1940b). The exhibition was held in St. Mary's Guildhall by the recently formed local division of the Association of Architects, Surveyors and Technical Assistants (consisting largely of members of Gibson's office). As Johnson-Marshall – the branch secretary – bluntly confessed, it was a propaganda exercise and intended to raise public awareness of the need for a coordinated approach to planning (Tiratsoo, 1990). A lavishly produced accompanying booklet described the purpose of the exhibition:

> . . . [T]he lines of this City of Tomorrow must be drawn up. Drawings and models must be prepared, so that every citizen can see this dream materialised and can say 'This is what I want, and this is what I will strive to have'. It is relatively easy and inexpensive to plan in this way, and it is work therefore which can and should be commenced now.
>
> (Johnson-Marshall collection, University of Edinburgh,
> GB 237/PJM/ABT/E, page unknown)

Around 5,000 people attended the exhibition (Tiratsoo, 1990, page 9), and visitors could record their comments on their experience of the exhibition using a visitors' book, which, unusually, survives. Most comments recorded were distinctly positive towards the redevelopment of the city centre even before the bombing provided the impetus for rebuilding (Johnson-Marshall collection, University of Edinburgh; comments summarised in Larkham and Lilley, 2012).

Some distinctly modern and 'technocentric' approaches were used to convey Coventry's case for redevelopment: the need for improved health, housing, education, transport, civic spaces and industry was made using detailed statistics and photographs mounted on panels. The slogan 'The Idea/To Avoid Chaos we must PLAN our city for all our needs' emphasised the possibilities of a 'new' Coventry, models of modern town-planning schemes were borrowed from the Architectural Association, from the Stratford-upon-Avon-based F.R.S. Yorke and Marcel Breuer and the 'MARS' group (Larkham and Lilley, 2012). Modernist architects including Maxwell Fry and Leslie Martin also loaned models of recently designed school buildings, and the RIBA sent a photographic exhibition of small houses. The centrepiece of the exhibition was the model, albeit rather crude, of the proposed civic centre made by the members of Coventry City Architect's department (Campbell, 2004, 2006). In terms of public response to the exhibition, the local press was largely enthusiastic: the *Coventry Herald* praised the City Architect's efforts to harmonise with surroundings and thanked him for his efforts in communicating the ideas of reconstruction to the people (Johnson-Marshall, cited in the *Coventry Herald*, May 1940a). There were some dissenters, though:

> The city is far better left as it is, in all its terrible ruin, pile on pile of stones, then it should be rebuilt to this new concept.
>
> (Carran, 1941, page 124)

Gibson, like Manzoni, built his reputation for planning expertise through talks to local organisations, presentations to national and professional bodies and publications. Locally, he said:

> [W]e give a lot of lectures: we always try to get a member of the Council to take the chair at these lectures, so that they will understand what is being said. Our architects go and give lectures to the schools so that we shall be building up in the future a useful population who understand these things.
>
> (Gibson, 1947, page 405)

He gave a widely reported lecture on planning to the Royal Society of Arts in December 1940 (though it had been prepared before the major air raid). The tenor of his message assumed a deeper significance when he spoke of the possibility that 'like a forest fire the present evil might bring forth greater riches and beauty' (Gibson, 1940a, pages 1583–1584). This is the 'opportunity' discourse used by many professionals, cast in different words for a different audience. He wrote a Preface to a publication exploring 'the practical principles of successful town planning' (Gibson, 1941a) and spoke to the Architectural Association on similar matters (Gibson, 1941b). He became an Associate Member of the Town Planning Institute.

Concentrated bomb damage in Coventry

The extent of bomb damage caused by the *Luftwaffe* attacks on Coventry and Birmingham varied. In Coventry, while the targets were ostensibly the factories distributed across the city centre,[4] the congestion and the nature of the tight-packed buildings in the inner-city core ensured that damage was extensive. The impact of the 'great raids' of autumn 1940 and spring 1941 have been lucidly recalled in the account of Reverend G. W. Clitheroe (of Trinity Church, Coventry) (1942) and described by John Petty (Provost of Coventry Cathedral during the 1980s and 1990s) (1987) '*Coventry Cathedral: After the Flames*'. The raid of November 14th, 1940 was vividly described by Clitheroe (1942, page 23) as: 'The raid began with a shower of incendiaries . . . the sound of water pouring on the flames could sometimes be heard, together with the burst of bombs, the swish of falling incendiaries and the fire of our guns'. In his recent reflection on the Coventry blitz, McGrory (2015) notes how Coventrians continue to live with the results and consequences of the raids, when the city was pummelled by bombs and was shaped, and still is shaped, by councils and architects who looked to respond to the opportunity afforded by the wartime destruction (see also Taylor, 2015).

The wartime destruction in central Coventry put earlier attempts of reconstruction into sharp relief. The main raid, of 14 November 1940 destroyed much of the city centre: this involved the destruction of more than 4,000 houses (NA CAB 87/11), together with 702,600 square feet

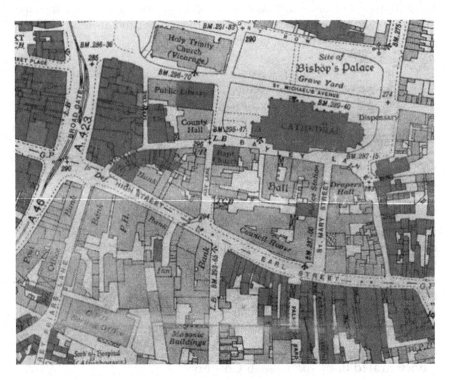

Figure 3.3 The extent of bomb damage in Coventry city centre
Source: Image copyright of Coventry Archive

of rateable retail and business use (Rigby-Childs and Boyne, 1953, pages 41–52) (Figure 3.3). Though there is some uncertainty about the estimated levels of destruction, Gregory (1973) suggested that 554 people lost their lives in that raid, which also resulted in the damage of several public buildings; the Cathedral, central library, swimming baths and a cinema were all affected.[5] The city's half-timbered buildings were especially susceptible to incendiaries, though Ford's Hospital, Bablake School and Bond's Hospital were only partly damaged. Holy Trinity Church survived, as did the Council House and St. Mary's Hall. Raids in April 1941 inflicted further damage on the centre.

The City Council's existing plans for reconstruction, therefore, encouraged the city's public to focus on looking to the future. Moreover, the suddenness and scale of the city's destruction assumed a different level of local and national importance to that of other British cities (Hodgkinson, 1970). Press coverage about the destruction helped transform the city into one of national and international significance and thus a source of powerful propaganda. That said, the City Council's reconstruction ambitions also stimulated disagreements between tiers of local and national government and amongst local commercial interests (Hasegawa, 1999); the path to post-war reconstruction was therefore far from straightforward.

Gibson is usually linked with the long-serving City Engineer, Ernest Ford (1884–1955), in most narratives of Coventry's reconstruction. Ford was a more traditional senior officer, appointed in 1924, with better links with the previous city administration than with the recently elected Labour politicians. He was responsible for a number of civic improvement projects though, with limited political or financial support, 'he was only able to make piecemeal adjustments to the street plan in the city centre when opportunities arose' (Gould and Gould, 2016, page 4). He did have a vision for the city centre before 1939, but it was less radical, retaining more existing alignments and properties, than that of Gibson.

Gibson's department was vital to the success of Gibson himself and his plans. As Johnson-Marshall (1958, page 225) noted, 'a few of us went to help start the new office, and we went bursting with ideas, ideas about prefabrication in building, about new kinds of housing layout, about carrying good design into every detail of the townscape and about making the whole city a collective work of art'. The radical young staff whom he appointed and who became a diaspora of Coventry-nurtured expertise, included John Barker (who became Bedfordshire County Architect), Brian Bunch (Deputy Architect and Planner for Stevenage New Town), Wilfred Burns (Newcastle City Planning Officer, later Deputy Secretary of the Department of the Environment and knighted, President of the Royal Town Planning Institute), Percy Johnson-Marshall (Professor of Urban Design and Regional Planning, University of Edinburgh), David Percival (Norwich City Architect) and Fred Pooley (County Architect for Buckinghamshire, then GLC Controller of Transport and Planning and RIBA President).

Dispersed damage in Birmingham

Birmingham also suffered several substantial (over 100 aircraft) raids, though several smaller raids between August 1940 and April 1941 also caused considerable damage (Larkham, 2016). By the end of July 1941, when the main raids had ceased, Birmingham had received eight major raids; by the end of the war, it had received nearly 2,000 tons of high explosive and incendiary bombs. Notwithstanding the imposed and voluntary censorship of Birmingham newspapers for morale and intelligence reasons (see Sutcliffe, undated), it was reported that, in total, 2,227 people lost their lives and 3,000 were injured (*Sunday Mercury*, 11 Nov, 1990, cited in Larkham, 2007). Birmingham was thus, equal with Liverpool/Birkenhead, 'the worst bombed city area in Britain after the capital' (Sutcliffe, undated, page 24, citing Young [1966] page, 54). The voluntary censorship was counterproductive in terms of public morale; it was felt

Figure 3.4 The 'Big Top' site under construction in the mid-to-late 1950s, New Street, Birmingham

Source: Reproduced with permission of Birmingham History Forum

that the city's suffering was being underplayed in comparison to that of Coventry (Larkham, 2007).

It is suggested that 99 factories (including a raid in November 1940 on the Austin factory, at Longbridge, which produced Stirling bombers for the war effort) (Sutcliffe, undated), nearly 4,000 business premises and 12,391 houses were all destroyed by the end of the war (Black, 1957).[6] Even so, more than 76 sites over half an acre were considered to be 'devastated' (Public Works Committee report, City Council minutes, 25 July, 1944, page 480). The two major inner-city areas affected by the raid carried out on the 9 April 1941 were the Bull Ring and the Market Hall and the corner of New Street and High Street later known colloquially as the 'Big Top' site (Figure 3.4). Although the damage was extensive in Birmingham, it was scattered and far less than suffered by Coventry and by many other European towns later in the war (Sutcliffe, undated, pages 25–26) (Figure 3.5). Direct bomb damage was relatively limited, to the extent that the City Council suggested that 'it was possible to walk through some parts of the city, the south-west and the north and see no signs of a recent German visit' (Birmingham City Council, 1995, page 5). Despite the weight of bombs falling

Figure 3.5 The extent of bomb damage in Birmingham city centre
Source: Reproduced with permission of the Library of Birmingham

on Birmingham, the city did not qualify for special government approval for replacement retail and commercial development (Larkham, 2016), because the damage was so scattered. Nevertheless, it was one of the 'test cases' studied by the Ministry in 1941.

Towards a 'Coventry of the Future'

Reflecting on his experiences of wartime Coventry, Gibson provided an evocative, if perhaps unsympathetic, description of how he and his staff would 'go up and see which buildings were [destroyed or damaged] to see how it would speed up the planning of Coventry' (Gibson, interviewed by Andrew Saint, NA, Architects' Lives C447/11, 1984). Gibson could take some comfort in having a department comprised of 'young radical' archi-tects, noted for its forward-thinking and collaborative approach to push-ing on with redevelopment. However, having to defer to the City Engineer (Ernest Ford) in planning matters was a hindrance to Gibson's belief that he should be 'the chief authority in all matters relating to the design, structure and appearance of the city' (Gibson, 1940b, page 3).

The association between the two men was, perhaps understandably, not entirely harmonious.[7] One of the strongest reasons underpinning their fractious relationship was the reciprocal respect that existed between Gibson and Alderman Hodgkinson, the chair of the City Redevelopment Committee. The committee was established in November 1940 (following the air raid) to produce a 'unified direction of the process of comprehen-sive redevelopment' (Gregory, 1973, page 96). The Town Clerk, Frederick Smith, approached central government for enabling legislation and assis-tance to implement plans for reconstruction. Both Gibson and Ford pre-pared plans for reconstruction. Gibson proceeded with his more radical plans for rebuilding. Ford's plan was to rebuild the city centre in a more conservative fashion, which he had been thinking about for several years.

A Coventry deputation, consisting of the Mayor and the Chairman of the Redevelopment Committee, was invited to London in January 1941 to meet Sir John Reith, the recently appointed Minister of Works and Buildings and First Commissioner of Works (Hasegawa, 1999). Reith offered the follow-ing advice for badly damaged Coventry:

> I told them that if I were in their position I would plan boldly and com-prehensively, and that I would not at that stage worry about finance or local boundaries. They had not expected such advice. . . . But it was what they wanted, and it put new heart into them.
>
> (Reith, 1949, page 424)

Immediate concerns following the bombing raids centred on the need to retain some sense of public order, but once this was restored, the desire to redevelop 'boldly and comprehensively' (following Reith's instruction) emerged as a new maxim for the City Council – as for other towns includ-ing Plymouth's 1943 plan (Hasegawa, 1992). In his autobiography, *Into the wind*, Reith suggested that 'Coventry would be a test case – not for me but for my government and for England' (Reith, 1949; see also Pevsner, 1951; Richards, 1952).

Seemingly, this sense of opportunity was embraced by the City Council, with the Town Clerk, Frederick Smith, acknowledging the 'helpful and sym-pathetic attitude displayed by Reith': 'Coventry (would) be given an oppor-tunity, such as never before occurred, to give expression to the aspirations of its citizens and civic representatives for a modern and beautiful city' (CCA Coventry City Council, 23 Jan, 1941). In February 1941 both Gibson and Ford's reconstruction schemes went before the Redevelopment Committee; Ernest Ford argued that:

[The] redevelopment scheme . . . should preserve the historic growth of the streets . . . old streets such as Cow Lane, Greyfriars Lane, Warwick Lane, can be conveniently used as backways . . . any ultimate plan which ignores the city's growth would be a national calamity.

(CCA, Reconstruction Committee Minutes 14 Feb, 1941)

Gibson suggested that the November 1940 raids had presented an opportunity to rebuild a city centre that the people of Coventry 'will enjoy living in instead of trying to escape from at week-ends' (Gibson cited in the *Daily Mail*, 24 Feb, 1941). Gibson was less than sanguine over Ernest Ford's scheme, stating that such a plan would fail to provide sufficient opportunities for 'architectural effects and for the amenities which our civilisation has given us the right to expect' (*Architects' Journal*, 24 Apr, 1941, pages 277–281).

In his personal testimony, Gibson explains how important his fellow architect, Percy Johnson-Marshall, was in shaping his ideas. Gibson explains the delicate task of drawing-up his vision(s) while maintaining a sometimes-strained working relationship with Ford, the City Engineer (Gibson, interviewed by Andrew Saint, NA, Architects' Lives C447/11, 1984). Gibson's 1941 (Figure 3.6) plan set out the principles and basic form of the city centre redevelopment that was carried through a whole series of iterations. It set out zones of activities, which bear resemblance to the ideas of Le Corbusier, Gropius and other European modernists, but it also reflected Abercrombie's ideas of regional plans published between the wars. In addition, Gibson drew inspiration from Abercrombie's use of ring roads to divert traffic away from over-crowded centres, and the grouping of buildings in 'precincts', though the term was promoted by H. Alker Tripp, assistant commissioner to the Metropolitan Police, in his *Town Planning and Road Traffic* in 1942.

Apparently unconcerned with the city's medieval street pattern, Gibson displayed a staunchly anti-sentimentalist approach to the replanning of the city centre, and his scheme all but removed the existing urban grain (see Figure 3.7). Opposed to the retention of surviving historic buildings, where these interfered with the plan, his first proposals swept away all the buildings on High Street, including the recently built National Provincial and Lloyds Banks. There were other suggestions for the removal of the badly damaged Ford's Hospital (Figure 3.8) to the Bond's Hospital site adjacent to Spon Street (although the map shows it remaining awkwardly on a small island in a new major traffic road):

Care has been taken to avoid disturbing ancient monuments, and wherever possible these have been incorporated as features or gardens. . . . Ford's Hospital could either be reconstructed on its present site, or it could be reconstructed in conjunction with Bond's Hospital [Bablake Street], forming a medieval group of interesting buildings.

(CCA, Coventry City Council Minutes, 17 Mar, 1941, page 77)

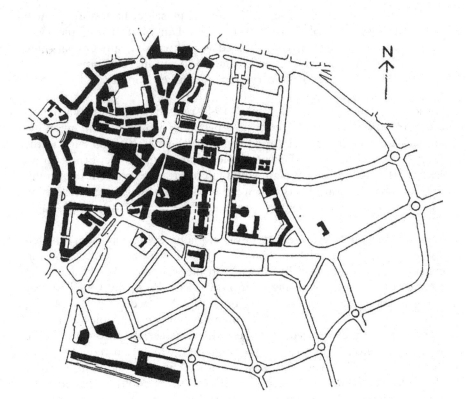

Figure 3.6 Ford's plan, January 1941
Source: Image copyright of Coventry Archive

Gibson was later called 'a malignant [who] has paid no attention at all to ancient monuments because he dislikes them' by the Government's Chief Inspector of Ancient Monuments (memo, 4 Dec, 1952, TNA WORK 14/1781).

The major shopping streets were to be removed and replaced by a sizeable pedestrian shopping precinct, liberated from the extraneous dangers of vehicular traffic and congestion, with the central axis aligned on the spire of St Michael's Cathedral. There were zones for entertainment and commerce situated to the west of the Cathedral and a large public space replete with cultural and legal assets located to the east (*Architects' Journal*, 24 Apr, 1941). Ultimately, the Council was swayed by Gibson's ideas: he was appointed as Joint Planning Officer for the City. This was a situation which, as Redknap (2004) states, must have proven deeply unsettling for Ford. The principles enshrined in Gibson's plan of 1941 established the fundamental design of the city centre reconstruction, and these were largely realised despite subsequent revisions to the plan (Gould and Gould, 2016).

Figure 3.7 Gibson's plan, February 1941
Source: Image copyright of Coventry Archive

Figure 3.8 Bomb-damaged Ford's Hospital (right); the building on the left was removed
Source: Reproduced with permission of Rob Orland

One of the most innovative features of Gibson's plan was reserved for Coventry's planned shopping precinct complete with rear service access and car parking, offering a 'purely pedestrian' (CCA, Coventry City Council Minutes, 25 Feb, 1941) experience away from the dangers of vehicular traffic. The design for the precinct also included 'two-level shopping rows', which, according to Johnson-Marshall (1966), was directly inspired by the design of the thirteenth-and fourteenth-century Chester Rows; this sought to create, notionally at least, a denser shopping provision with higher rate-able values.

There were several key features of the Gibson plan. It centred on the need to improve traffic circulation and road capacity within the city centre; to 'de-clutter' and separate out individual building units; to cluster together buildings catering for the same form of activity – shopping, administration, recreation, business (*Architects' Journal*, 24 Apr, 1941). The importance of main circulatory routes, grouping of activities and proportion of open space to buildings was also emphasised (*Architects' Journal*, 24 Apr, 1941). Gibson also proposed an inner circulatory road that followed the line of the existing Queen Victoria Road, Corporation Street, Ford and Lower Ford Streets and Vecqueray Street, looped south to Acacia Avenue and ran west, close to the railway station rejoining Queen Victoria Road at Greyfriars Green. Ford's plan was essentially more conservative, without the shopping precinct and retaining more traditional road alignments. However, he did suggest a substantial civic precinct south of the cathedral. The City Council asked both to work together and, eventually, a compromise plan was accepted; although Gibson's ideas were clearly dominant.

During late 1941, a semblance of familiarity resumed in the city centre with the erection of specially designed temporary shops in Broadgate (*Architectural Review*, Oct, 1941, page 110). The bomb-damaged Cathedral, however, proved to be more problematic. The local newspaper magnate, Lord Iliffe, offered to pay for a large model of the city centre scheme and the Cathedral, as the centrepiece of the design, had to be shown, but questions were raised over as to whether it should be presented as ruins, fully reconstructed, or as an entirely new building. The model was completed in 1944. The new Cathedral, designed in 1944 by Sir Giles Gilbert Scott (the designer of Liverpool Anglican Cathedral), had its nave running north-south, connected the old spire through a cloister formed from the old walls. But Scott disagreed with the Bishop and withdrew. Ultimately, Basil Spence won the competition to design the new Cathedral, and his text, *Phoenix at Coventry* (Spence, 1962) which describes the rebuilding of the Cathedral, evokes the idea of the city's symbolic rebirth after destruction (Campbell, 2006). The model showed a remodelled ring road as a dual carriageway at grade with seven roundabout connections, while the buildings were designed to be five to seven storeys high.

Gibson's inspirations

Early in 1941, Gibson's *Plan for the New Coventry* was reprinted in the *Architect and Building News* (1941). Subtitled *Disorder and Destruction: Order and Design*, this plan juxtaposed perspective drawings showing the 'new Coventry' with an aerial photograph of the pre-war city, arguing to readers that 'this must not happen again' (*Architect and Building News*, 1941, page 192). Here there was a palpable anxiety regarding the city's 'failings'. They were listed as 'narrow roads, chaotically placed; an irrational mixture of shopping, industrial and residential areas; an absence of large, green spaces which would give the city a chance to breathe; large slum areas; and the ugly and depressing pompous foolery of architecture, killing its very soul' (CCA Coventry City Council Minutes, 1941, page 192). Positioned against the background of wartime solemnity, Gibson questioned:

> Can we afford to cease to work for a creative end, and even though we are at war? If we do not, the open gate of defeat lies ahead, and behind it the declining path of civilisation and decadence. . . . Here is the challenge! If we do not take it up we are indeed decadent, defeated.
>
> (CCA, Coventry City Council Minutes, 1941, page 195)

There are strong and obvious parallels between Gibson's rhetoric and that of Le Corbusier, who spoke of the chaotic effects of uncontrolled, disorderly urban growth – 'the miasmas of anxiety now darkening our lives' – associated with the evolution of 'great' cities:

> [W]e may admit at once that in the last hundred years a sudden, chaotic and sweeping invasion, unforeseen and overwhelming, has descended upon the great city; we have been caught up in this, with all its baffling consequences, with the result that we have stood still and done nothing.
>
> (Le Corbusier, 1929, page 25)

Johnson-Marshall acknowledged that aspects of Le Corbusier's vision of the contemporary orderly city influenced the ideas coming from the City Architect's Department (cf. Johnson-Marshall, 1966, chapter 6). It would be a mistake, however, to assume that European modernist ideas were slavishly followed during the rebuilding of Coventry (Campbell, 2006). There were also shades of more moderate modernism in Gibson's thinking (Tiratsoo *et al.*, 2002; Campbell, 2004). Significantly, perhaps, Gibson admitted that in central Coventry, there was no compelling argument for building high skyscrapers; vertical accents would be provided by the spires of Coventry's three great medieval churches[8] and not by the towers of a business quarter, as in Corbusier's *Plan Voisin* (Johnson-Marshall, 1966). Moreover, the symmetrical geometry of the precinct emphasised the importance of the Cathedral spire; this approach shares some similarity with the *Beaux-Arts* approach of architectural planning, still evident in a few of the reconstruction plans such as that for Plymouth by Paton Watson and Abercrombie (1943).

A more balanced interpretation, perhaps, would suggest that Gibson was a pragmatist who was prepared to modify his ideas in the light of experience, events and local opinion: someone whose planning concepts perhaps owed more to Patrick Abercrombie and Lewis Mumford than to the precepts of the *Congres Internationaux d' Architecture Moderne* (Campbell, 2006). Mumford's book *The culture of cities* 'had recently arrived in England, and a colleague had straight away sent it up from London. We thought it so important that we passed it on to our Councillors to read' (Johnson-Marshall, 1966, page 295). In the words of Mumford, the industrial city was nothing less than 'the crystallisation of chaos': a world defined by 'a raw, dissolute environment and a narrow, constricted and baffled social life' (Mumford, 1938, page 7–8). The need for a cultural heart for Coventry was reflected in official discourse of the time, including Gibson's personal testimony, when he suggested that 'I want to make Coventry a city which people enjoy living in instead of trying to escape at weekends' (*Daily Mail*, 24 Feb, 1941):

> It has been said that Coventry is not a city of 'culture', and the reason given by some for this lack is bound up in the lack of a 'Cultural Centre' in the City. This is undoubtedly a real need. While elementary schools are not affected by any scheme for the re-development of the city centre, the establishment of a University College is envisaged. An adequate Art School is a necessity and its situation near the Art Gallery would be fitting. Into this centre would naturally come the Library, modernised and extended to provide in addition to library services, a hall of rooms for the accommodation of meetings of cultural societies, etc.
>
> (CCA Coventry City Council notes
> of redevelopment, 23 Jan, 1941)

Here it was expected that the inhabitants of a rebuilt city might begin to comprehend the nature of citizenship. Gibson, for example, spoke of the need for open space encircling key civic buildings:

> The fine Cathedral spire, Trinity Church and the medieval St. Mary's Hall form a group worthy of a fine display. An open space has, therefore, been planned around them, and it should again appear as it did. . . . It is suggested that this central area, which originally contained several factories, should be opened out as a small central park, in which new

civic buildings could be . . . grouped, and from which magnificent views of the Cathedral and the new buildings can be obtained.

> (CCA Coventry City Council,
> Gibson, 17 Mar, 1941, page 77)

Against this, however, there were criticisms that such 'opening' of the city centre would involve the irrevocable destruction of parts of the medieval urban fabric, including aspects of its domestic, civil, or ecclesiastical antiquity:

> [Any plan] should be to conserve, so far as is possible . . . a scheme of rehabilitation on purely modernistic lines, formal and rigid, would be not only inappropriate but injurious.

> (W. Randolph, letter to the
> *Birmingham Post*, 28 Jan, 1941)

Problems of implementation

Despite the Coalition government's strident proclamations of 'bold' planning, there was in fact a rather protracted and strained relationship between the City Council and Whitehall – one determined to realise its plan, the other set on reducing its scale and ambition (Hasegawa, 1992, 1999). Hence Coventry's 1941 Plan, while apparently being acknowledged by both the city's residents and by central government, did not evolve in a particularly swift fashion. Despite supporting the scheme, the cautious Town Clerk (Smith) and Treasurer (Larkin), sought assurances as to how it was to be financed and enabled (Hasegawa, 2013), while existing local planning legislation proved burdensome and unyielding (Redknap, 2004). Moreover, central government was also sluggish to push through the legislation that would accelerate the rebuilding process – the 1944 and 1947 Acts (Hasegawa, 2013).

Despite being one of the first blitzed cities and a 'test case' for post-war reconstruction, the situation in Coventry was overshadowed by subsequent events taking place elsewhere. Other cities were bombed, of course; and this altered the Government's stance on supporting their attempts to reconstruct. Consequently, there was an uneasy stalemate between central government, which urged Coventry to develop and submit its plans for approval and the City. Protracted discussions ensued between the City Council and the newly formed Ministry of Town and Country Planning (MTCP). Consent was needed for the design before compulsory purchase of land and development loans could be given under the terms of the 1944 Town and Country Planning Act. The Ministry delayed, objecting to the excessive area of the civic precinct, to the arcaded, pedestrianised shopping area and to the expense of the plan in general (TNA HLG 79/130). The Council's response was that these were 'essentially matters to be decided by the judgement of the local authority' (Coventry City Reconstruction Committee, 13 June, 1945). There were other government concerns regarding the fact that local business interests were not consulted. For instance, the Chamber of Commerce and Multiple Shops Federation called for direct vehicular access to the main (pedestrianised) shopping area (for a discussion of this debate, see Fischer and Larkham, 2018, page 91).

In fact, the Chamber of Commerce had in 1941 suggested a geometric semi-circular layout for the retail core; in 1943–1944 this was revived, with a plan drawn up by F.W. Woolworth's architects showing below-ground servicing and access (Ministry notes, TNA HLG 79/131) (Figure 3.9). By mid-1945, a compromise had been reached with the Chamber of Commerce and other stakeholders: a new trafficked cross street (the future Market Way and Smithford Way) bisected the pedestrian area (TNA HLG 79/131). The principal pedestrian route ran beneath it; parking was allowed along its length, suitably near the shops. Significantly, the ring road design became significantly more complicated, although it encircled an area slightly smaller than the original plan suggested (for a review of the ring road, see Holliday, 1973). Queen Victoria Road and Corporation Street in the west and Cox Street in the east were left as 'local traffic' roads serving the precincts. A new ring road ran outside of them and land inside the road on the west was zoned for 'service industries'. This road, though, was seen as a mistake by most of the planner/architects.

Overall, although Coventry caused problems by pressing ahead with plan-making, following Reith's exhortation, many of its problems were

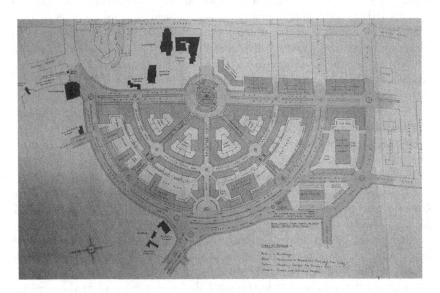

Figure 3.9 Chamber of Commerce plan 1944
Source: Image copyright of Coventry Archive

caused by government delays (with legislation) and prevarication (in examining and approving the plan). The new Ministry's new Planning Technique section was not supportive, although this is typical of a 'not invented here' mindset, critical of many plans and planners across the country – including those who had previously worked for the Ministry and its predecessors. The Council stuck to its guns and demonstrated a high level of political commitment to the radical plan. The government's initial support for radical replanning quickly waned. In opening the 1945 exhibition, the Mayor stated (in the Minister's presence) 'there has been all sorts of manoeuvring in order to edge Coventry away from the splendid designs it had' (quoted in *Midland Daily Telegraph* 9 Oct, 1945). Nevertheless, Coventry certainly had powerful chief officers, supported by local councillors, who were able to engage and mobilise the public in favour of their proposals (cf. Tiratsoo *et al.*, 2002; Hasegawa, 2013). Brian Redknap, the City Engineer (from 1974 to 1989, who joined the Council in 1956) argued that momentum was increased when the newly appointed Town Clerk, Charles Barratt and the City Treasurer, Marshall (both appointed in 1946) were introduced as the post-war replacements for the more conservative Larkin and Smith. Alongside local politician Alderman Hodgkinson, these individuals were particularly prominent figures in taking the opportunity to force through the adoption of a comprehensive city-centre reconstruction plan:

> [In terms of the re-planning] Ernest Ford really had his nose put of joint because there were a lot of public buildings to be designed and it was appropriate that this should be carried out by the city engineers' department we've got to have an architect and they [the Council] chose Gibson. . . . He came, and he produced a plan for the central area – he started on his own initiative and he got the ear of Alderman Hodgkinson and the two got together and came up with a plan for the redevelopment of [the city centre]. . . . I got to know Hodgkinson well – he thought, rightly in a way, that his . . . greatest achievement was to back Gibson's ideas.
>
> (Brian Redknap, interview)[9]

Birmingham's different approach

Many towns undertaking a reconstruction and replanning process at this time, whether bomb-damaged or not, engaged external consultants or had used their own in-house staff to produce a plan, for the town centre at least. Birmingham pursued an altogether different route towards rebuilding the city, deciding instead to eschew a city-centre comprehensive plan in favour of a more 'piecemeal' approach. Here, the influence of Herbert Manzoni and other professional officers, the elected members acting as Committee chairs and the developers and their architects was significant in shaping the post-war replanning (Larkham, 2016). Manzoni's major contribution to the city came with the realisation of two closely linked schemes, the completion of which was largely due to his personal determination and the high regard in which he was held by local politicians (Sutcliffe, 1967–1969). The first was a network of ring roads around the city core. The second was for five giant slum clearance areas, identified by 1941, which were to be comprehensively demolished and rebuilt in a modernist style (Sutcliffe and Smith, 1974). By the early 1970s, the ring roads were almost completed and the five slum clearance areas, totalling 563 hectares, had been completely rebuilt, with almost none of the original buildings surviving (P. Jones, 2004).

The city did, as indeed did many cities, including Coventry, appoint a 'reconstruction committee'; although in practice its influence on physical urban reconstruction was minimal. It also appointed advisory groups including representatives from interested organisations outside the Council itself. In late 1941, by which time there had been several months without further major bombing raids, the city's Public Works Committee (PWC) considered the issue of reconstruction. An internal report discussed the Civic Centre and the 'redevelopment areas', although all of these schemes had been identified and, to a greater or lesser extent, planned before the war: the issue of war damage was not raised (copy bound with PWC Minutes, 23 Oct, 1941).

There is some divergence between the approach to reconstruction taken by Gibson and Manzoni. While Gibson displayed a rather radical view on reconstruction, Manzoni argued, in a short interview to the *Birmingham Mail* in early 1941 (27 Feb, 1941), that Birmingham had no such grandiose plans for rebuilding, suggesting instead that 'all we want is the opportunity to carry out the plans we have already'. For him, the dispersed nature of the bomb damage ensured that any extravagant ideas that a 'new city can emerge, Phoenix-like . . . is quite erroneous' (*Birmingham Mail*, 27 Feb, 1941). In contrast to the seemingly dominant view that the bombing raids presented an 'opportunity' for comprehensive replanning (e.g. Tubbs, 1942, page 21), Manzoni presented a more realistic perspective: for him, reconstruction plans 'were often obsolete by the time they were put into effect'. This was a view reinforced by his experiences of the earlier failed proposals for the grandiose civic centre, dating from the First World War (Sutcliffe and Smith, 1974, page 448). It is perhaps surprising, therefore, that in October 1941 the Public Works Committee for Birmingham put forward the idea that a comprehensive approach was necessary to improve the city:

> The incidence of damage caused by air raids is an immediate consideration. [It gives] an opportunity to affect many improvements of a local character which in themselves bring near the fulfilment of more comprehensive schemes of redesign and the general awakening of the public conscience will probably demand that every opportunity be taken to make good the defects of the past.
>
> (PWC Minutes, 23 Oct, 1941)

For Manzoni, long-term plans for the city centre had to be 'flexible as possible', allowing for 'changes in taste and planning and architectural techniques' otherwise he feared that the plan 'will never be carried out at all' (Manzoni, 1968, page 4). Manzoni argued in April 1942 that 'the reconstruction of our towns and cities, whether bomb damaged or not . . . is an undertaking which calls for the collaboration of several professions and consultation with many others' (Manzoni, 1942, page 36). He went on to argue that engineers – rather than architects – *à la* Gibson – should be central to the process of rebuilding.

It is also arguable that Birmingham was in far stronger position than Coventry to proceed with reconstruction because of the range of plans for zoning (from 1913) and road plans (of 1919) that had already been well developed *before* the onset of the Second World War. In addition, Manzoni in had personal involvement with the shaping and drafting of the 1944 'Blitz and Blight' Town and Country Planning Act (and other Government guidance) – legislation that effectively initiated the prospect of wholesale comprehensive redevelopment. The provisions made within this Act enabled compulsory acquisition, thus helping pave the way for redevelopment in the city (Sutcliffe and Smith, 1974). He suggested in a 1967 interview with Anthony Sutcliffe that:

> I was asked by the government to go on a small committee to consider the rebuilding of towns which had been damaged by enemy bombs. . . . We tried to talk out the problems of rebuilding these damaged cities. . . . I can't say that I initiated it [but it ensured that] by a simple enquiry we could get compulsory purchase on a vast area.
>
> (Manzoni, interviewed by Sutcliffe, 1967–1969, page 3)

When this legislation was passed, Manzoni commented that 'we in Birmingham are ready, because our plans had already been drawn up in detail, and we took advantage of these powers' to acquire the five development areas (Manzoni, 1968, page 2). Manzoni could not act alone, though and he noted that 'the influence of officialdom in town-planning has been very considerable', and certainly, there were other powerful personalities that helped shape Birmingham's post-war development.

Frank Price was instrumental in promoting development, attracting prospective developers – including, until they were taken over, Harrod's – and levering funding from the central government, especially for the inner ring road. Changes made at the administration of local government are significant, too. When the Unionist party won control of the Council in 1949, the Council countenanced changes to the administrative structure, which would relieve Herbert Manzoni of his architectural responsibilities (Sutcliffe and Smith, 1974, chapter XIII). Alwyn Sheppard Fidler (1909–1990) was appointed, and he took over some of the work undertaken by Manzoni, with an apparently independent Department. Sadly, however, according to

Sutcliffe and Smith (1974, chapter XIII), Fidler was not consulted on the overall development of the city centre proposals; he felt subservient to the workings of the powerful Public Works Department, which, in his view, was orchestrating proceedings.

In contrast to the concentrated bomb damage of Coventry, it is difficult to present a coherent narrative of post-war redevelopment in a city as large as Birmingham, where the severity of wartime destruction was tempered by its dispersed nature and where no single rational reconstruction plan was ever formulated. Whereas the plans for Coventry's redevelopment changed, Gibson's comprehensive approach was largely 'pushed through', buoyed in part by professional and public support, the narratives of the Birmingham-trained and based architects James Roberts and John Madin provide a largely critical view of what they perceived to be Manzoni's lack of foresight in failing to pursue a reconstruction plan for the city centre:

> There's interesting stories about Manzoni because ultimately when I came back from the army I was very keen on comprehensive planning and I suggested to Manzoni that there were only about three freehold interests [within the city centre] and what he should do is do a comprehensive plan for the whole of the centre of Birmingham within the ring road. But I thought here was a great opportunity. The city itself owned quite a lot within the ring road and I thought this was a great opportunity to produce a plan. . . . But he didn't go along with this.
>
> (Madin, interview)

A similar line of argument has been recalled in other accounts: 'I said that there should be a three-dimensional plan, but he replied that it could not be done. . . . So much of the land in the city centre was controlled or owned by the Corporation. But all that we architects have been allowed to do is to plan parts of the city centre. We are only pawns in a very big chessboard. . . . Basically, Manzoni was road engineer. He was a fine chap and a great friend, but it was his limitation' (Madin, interviewed in Sutcliffe 1967–1969). This may reflect editorial priorities as much as public views. James Roberts also spoke of Manzoni's 'unsympathetic', forthright and 'uncoordinated' approach to the replanning of Birmingham:

> Manzoni I got to know very well but . . . he had no interest in architecture at all, [no interest in] aesthetics at all, he wasn't interested in people or pedestrians. It was cars, lorries, getting things through and out again and so he did considerable damage to the heart of Birmingham I think but there should have been a lot of tender loving care after the war. I remember the first streets and areas were in Union Street in Birmingham just about where the present inverted pyramid which John Madin did [for the Central Library], there was, it was Union Street, but it was, it was, there were a couple of very lovely pubs and which we

would have kept as fresh as 'Wow they've survived'. But also, on the corner opposite the council house there was a most wonderful Gothic building which survived all the bombing and that was the reference library. It was a huge great big building where you went up a set of stairs with wooden insets in them and handrails but that was a building. Mr Manzoni knocked it down after the war; it escaped the war, [but] he knocked it down because he put the ring road underneath it, the tunnel underneath it, so it came down.

(Roberts, interview)

Manzoni's influences

In Manzoni's own words, '[a]s to Birmingham's buildings, there is little of real worth in our architecture. Its replacement should be an improvement. . . . As for future generations, I think they will be better occupied in applying their thoughts and energies to forging ahead' (Manzoni, source unknown, quoted in Foster, 2005). Implicit within Manzoni's rhetoric was the belief that he was unconcerned about architectural style, being more preoccupied with pushing forward and using standardised house designs often drawn up by building contractors. It should be noted that Manzoni was being politically pushed for completions and constrained by budgets and that there was also the issue of pressure for high densities, influenced by the scarcity of readily developable land within the city and the high cost of agricultural land outside it.

As with other bomb-damaged towns, the pressing need for redevelopment of the city at that time was apparently self-evident: Birmingham was booming on the back of the motor industry, with rapid suburban expansion in creating problems of traffic congestion and urban blight in the inner core (Cherry, 1994). The full Council subsequently declared a 267-acre area as a Redevelopment Area under Section 34 of the Housing Act 1936 (Birmingham City Council Minutes, 14 Dec, 1937)[10] – the largest such scheme anywhere in the country at the time (Cherry, 1994).

On the 4 July 1939, the city decided to prepare a similar plan for 12,600 acres of the city's central area, but Ministerial approval was delayed by the onset of war (NA HLG 79/932; PWC Minutes, 4 July, 1939). By 1941, the early plans for Duddeston and Nechells were joined by plans for four more 'redevelopment areas', together forming a massive programme to rebuild a ring of decay around the city centre (Sutcliffe and Smith, 1974). The five reconstruction areas already identified as a focus for slum clearance and rebuilding (see Sutcliffe, 1967–1969). The plan required for the redevelopment areas identified the broad layout and land use, though the plan was finally presented to the Public Works Committee on 27th May 1943 (Manzoni, 1943).

As Glendinning and Muthesius (1994) point out, the 1943 model for Duddeston and Nechells shows a line of towers bearing a striking resemblance to the Soviet Pavilion at the 1937 World's Fair. Overall, however, this was a curious assemblage of different architectural styles, at this stage perhaps modernistic rather than modern, but all suggesting a forward-looking city wanting to efface elements of the existing urban landscape (P. Jones, 2004) (Figure 3.10).

Proposals were put forward in 1945 for Duddeston and Nechells, Summer Lane, Ladywood, Bath Row and Gooch Street, totalling 1,382 acres (Larkham, 2016). These areas were often referred to as the five 'New Towns' and were meant to be comprehensively developed communities, with new housing, schools, shops and other public facilities. In 1958, these areas became Nechells Green, Newtown, Ladywood, Lee Bank and Highgate after a public competition (Price, 2002, page 197). The underlying rationale behind these proposals is perfectly elucidated by a 1957 report, published by the Council, which wrote of its 'intention to create . . . a balanced community' and quoted Patrick Geddes' 'interpretation of the aim of town planning to create a fuller, greater and happier life for a city's inhabitants' (Birmingham City Council, 1957). A new Central Areas Management Committee between 1946 and 1951 managed the redevelopment areas. Within the five areas, some 30,000 houses were demolished; this was necessarily a phased process and, having acquired all sites via compulsory purchase, the city found itself managing thousands

Figure 3.10 Duddeston and Nechells Redevelopment Scheme
Source: Reproduced with permission of the Birmingham Museums

of additional properties that had to be repaired, tenanted and managed before their demolition was scheduled (Glendinning and Muthesius, 1994). In the 'new towns', each area was zoned into different defined functions: a substantial amount of public open space was incorporated and through traffic restricted to certain routes rebuilt to a high standard (Higgott, 2000) (Figure 3.11).

The first physical reconstruction was in 1951 in Duddeston, and it became a showpiece to illustrate how a new Birmingham might take shape (Higgott, 2000). It originated with a proposal made in 1937, centred on the building of a dual carriageway 'Parkway' flanked by tower blocks of flats. According to Manzoni, 'practically the whole of the property is old and dilapidated (thus) the area is suitable for the redevelopment of a single and unified scheme' (Manzoni, 1943). The Council's intention was not only concerned with the number of flats built and the speed and efficiency of its programme, there was also the underlying belief in creating a new and better form of community. The whole area was the product of erasure, with nothing of the earlier drabness remaining, and the nineteenth century urban fabric had disappeared (P. Jones, 2008).

Figure 3.11 Redevelopment of Nechells Green
Source: Reproduced with permission of the Library of Birmingham

This story is substantiated by Manzoni's own testimony concerning reconstruction, and perhaps stung by the failure of the grand 1920s Civic Centre scheme, Manzoni spoke of his desire of looking forward:

> I have never been very certain as to the value of tangible links with the past. . . . As to Birmingham's buildings, there is little of real worth in our architecture. Its replacement should be an improvement, provided we keep a few monuments as museum pieces to past ages.
>
> (Manzoni in 1957 quoted by Foster, 2005, page 19)

Communication of planning ideas to the public

In October 1945 celebrations marking the end of wartime, the city's 600th centenary of the Incorporation of Coventry and the bright new future that had been promoted to its citizens coalesced in the 'Coventry of the Future' exhibition (Coventry City Council, 1945). Comprehensively covering themes of roads and transport infrastructure, to the position of housing and industry, spaces for recreation and public services, the exhibition included elaborately constructed displays and models to communicate to the public what the possibilities of living in their 'new' city (Larkham and Lilley, 2012) (Figure 3.12). Beginning with an introduction, which deliberately encouraged the visitor to consider the 'necessity to plan', the exhibition was also designed to persuade members of the public that effective planning was needed and sought the support of enlightened public opinion (Coventry City Council, 1945). Following a carefully structured path through the displays,

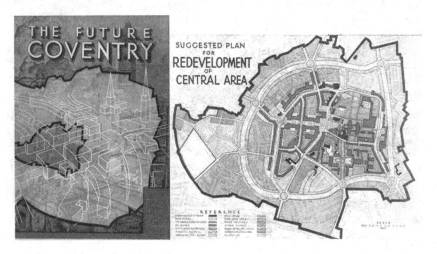

Figure 3.12 Coventry's 1945 Plan

Source: Image copyright of Coventry Archive

visitors were guided to the finale of the exhibition: the plans for city centre redevelopment and Lord Iliffe's model (Walford, 2009).

The exhibition, furnished with a café, and more than 57,500 visitors (Coventry City Council, 1945, page 3) experienced a cinema showing educative films on the importance of post-war planning, reconstruction and housing. Accompanied by a sleek, illustrated booklet, the exhibition provided details of the proposed civic quarter, new shopping precincts, municipal baths and an entertainment district. A competition sponsored by the Council in collaboration with a local newspaper, the *Coventry Evening Telegraph*, also invited 'suggestions' and 'to ascertain views of Coventry people as to general improvements' (CCA/TC/27/1/4).

Unlike Coventry, unofficial attempts were made to communicate Birmingham's approach to reconstruction. The Cadbury and Bournville Village Trust publication, *When We Build Again* (BVT, 1941) was influential in several respects, as it drew readers' attention to some of the unfortunate truths about the state of the city's housing and living conditions and broadly identified what might be done to alleviate such problems (Larkham, 2013). There were also radical proposals embedded in this publication – for example, the wholesale modernist-inspired redevelopment of the Jewellery Quarter, an area to the north of the city centre, which the City Council recently suggested was worthy of World Heritage status (Figure 3.13). Paul S Cadbury's lavishly illustrated volume *Birmingham – Fifty Years On*,

Figure 3.13 Jewellery Quarter with proposed flatted factories: perspective by J Schreiner

Source: Reproduced with permission of the Bournville Village Trust

published in 1952, purposefully juxtaposed photographic images of the eighteenth and early-nineteenth century Birmingham together with maps, images of street scenes and buildings and transport systems in an imagined city of 2002 (Cadbury, 1952) (Figure 3.14). Yet Manzoni contributed to the West Midlands Planning Group's 1948 publication, *Conurbation: A Planning Survey of Birmingham and the Black Country* (West Midlands Group, 1948). This document was remarkable not only in its depth and in breadth of research and use of thematic land-use maps, but also for its

Figure 3.14 New Street in 2002

Source: Reproduced with permission of the Bournville Village Trust

use of photographs taken of Birmingham and the West Midlands region. Frank Price made further efforts to communicate the Council's ideas in a special local newspaper supplement of 1959 entitled 'The New Birmingham' designed to: 'provide the citizens of Birmingham, with a record of the dynamic steps being taken to improve their Birmingham' (*Birmingham Evening Mail*, 1959, page unknown).

Writing some years later, Manzoni recalled that 'because slums weren't only slum buildings, they were also slum conditions of all sorts including a slum mentality . . . [if] we get rid of the lot, then we cannot only get rid of our slums, but we can make a new environment' (Manzoni, 25 Sept, 1967, quoted in Sutcliffe, 1967–1969). It could be reasonably argued, therefore, that such an attitude chimed with Le Corbusier's controversial ideas and the precepts associated with the clean-sweep approach that Le Corbusier promoted when he spoke of 'the need for the builder to bring order . . . for all around him, the forest is in disorder with its creepers' (Le Corbusier, 1929, page 71).

Summary

Chapter Three has explored the 'official' stories of the pre- and post-war context in which reconstruction ideas were conceived, refined and communicated to the public. This chapter has re-evaluated these ideas through constructing new narratives drawing on both original sources, (auto) biographical material of those involved in shaping the plans and more recent work on the two cities. Although this narrative of post-war change has been documented elsewhere, this chapter has sought to compare the different contexts and approaches to replanning in Coventry and Birmingham – this is a perspective that, to date, is underdeveloped in the existing literature on British post-war reconstruction planning. Chapter Four moves away from documenting and discussing the planning ideas developed in the two cities, to discuss how these concepts were experienced, concentrating more on the rebuilding activity in Coventry and Birmingham.

Notes

1 However, there is some dissent as to the origin and originator of these early ring road proposals: see Sutcliffe and Smith (1974, page 7).
2 Though the City Council owned many parcels of land in the city centre, wealthy landowners were still influential in shaping development decisions.
3 The West Midland Group on Post-War Reconstruction and Planning was formed in 1941. Early contributors to the group included: Dr R. E. Priestley, Vice-Chancellor of Birmingham University; Professor T. Bodkin, Barber Professor of Fine Arts at the University and Chairman of the Birmingham Civic Society; G. W. Cadbury of the Bournville Village Trust; P. S. Cadbury, Managing Director of Cadbury Bros; H. J. Manzoni, City Engineer and Surveyor of Birmingham City Council; and C. B. Parkes, Chief Architect to the Bournville Village Trust.

Membership was enlarged to include, for example, Dr L. Dudley Stamp, Director of the Land Utilisation Survey of Great Britain (see NA WMG 1 / n.d.)

4 For example, Daimler had a shadow factory at Capmartin Road (*c.* 1.2 miles north of the city centre), Rootes was located at Aldermoor Lane (*c.* 1 mile to the south-east of the city centre) and Standard had a factory at Fletchamstead Highway (*c.* 2 miles west of the city centre). Other factories located in the city, such as SS Cars (which became Jaguar after 1945), Alvis Motor Company, and GEC all manufactured products for the war effort (see Douglas, 1983).

5 Further *Luftwaffe* attacks ensued and in the April 1941 raid, some 475 people died with St Mary's Hall, the Police Station and Hospital all being damaged.

6 The figure for houses broadly matches the figure presented by Manzoni at a meeting of the Public Works Department in 1941 (PWC Minutes, 22.05.41).

7 However, professional and personal disagreements may have been over-emphasised.

8 Greyfriars had survived, although the rest of the church did not.

9 Redknap also suggested that "Charles Barratt [was also] an eminent figure in local government and very influential in processing the bureaucracy of redevelopment. . . . He was a person of national stature in Westminster and the treasurer who replaced Larkin was Dr Alfred Marshall and he was influential at national level [too]" (Brian Redknap, interview).

10 The printed Minutes of the full Council and its sub-committees are available in the Library of Birmingham.

4 Memories of rebuilding

Introduction

This chapter concentrates on the rebuilding activity in Coventry and Birmingham by drawing on original sources and interview material associated with those actors involved in giving shape to the post-war reconstruction of both cities. The ideas, concepts and visions for the redevelopment of both city centres were underpinned by two key planning principles. First, the safe and convenient segregation of pedestrians and motorised traffic facilitated by the construction of circulatory inner ring roads and the recommendations for the development of pedestrianised precincts; and second, the land-use segregation and the dedication of specific spaces to shopping, industry and recreational activities. These ideas were dominating planning thought in the 1940s, although their origins were earlier – partly influenced by US experiences, partly by the number of fatalities on increasingly congested roads.

Memories of the wartime destruction

Because of the concentrated nature of the bombing, Coventry had lost notable elements of its city centre; the old Broadgate and Smithford Street, the heart of the city's commercial heart, together with the Cathedral, which had been damaged beyond economic repair. Some respondents who contributed to Hubbard *et al.*'s (2003a, 2003b, 2004) study suggested that the 1940–1941 wartime raids on the city centre had left a deep mental impression:

> On the night of the blitz, there was a lull and we came out to see what was happening and the whole of Coventry was ringed with fire, it was a moonlit night and everything they could see – there were no doubts of where the car factories were – although they were camouflaged. . . . It was just a fire all around, it was dreadful. It went on all night. You could smell the burning the next day; we thought we'd better go to work. We hadn't been to sleep all night. And we went to the Alvis [motor car] factory and it had gone every bit of it. . . . The shelter I was in charge of was in Broad Lane, and I always remember there was [*sic*] some parked

cars parked in a ditch by the common and I almost crawled under one of those – I mean that was a silly thing to do to get under a parked car. I mean a direct hit a car with petrol in, you would know about it. I just can't believe we ever came out of it.

(Winifred, born 1915, interviewed 2001)

The November 14th blitz and the sky was lit up – it was just like the dawn coming up or sun going down. . . . You'd see the glow in the sky [it was] horrendous. And then the raid of April 8th the following year they did another go at Coventry and the April. It was horrendous [though we were alright]. We saw all the wreckage lying about Owen Owen's [on Broadgate] was burnt out. It was just a shell and all these other shops that were absolutely wrecked, you know, bomb damage all over the place. The Cathedral was just a shell. Hertford Street was a mess.

(Basil, born 1938, interviewed 2001)

Similar recollections came from those respondents who could remember the wartime destruction of central Birmingham. Birmingham's bomb damage was far less significant than other European towns – including Coventry. The two major inner-city areas were the Bull Ring and the Market Hall and the New Street/High Street corner later known as the 'Big Top' site. In April 1941 this site suffered 'severely, with huge fires burning in the Bull Ring, the High Street, New Street and Dale End' (Ray, 1996, page 255). For George, the damage to New Street held a particularly strong memory:

There was a big bomb in New Street by the Times . . . the furniture place . . . you know when you were that young, you didn't take much interest in the bombing to be honest. But I can remember the Bull Ring, the Bull Ring getting bombed and the fish market – the Market Hall. And . . . you could see Coventry getting bombed from Yardley. It was getting really bombed that was.

(George, born 1933, interviewed 2007)

The locally trained architect James Roberts also recalled how witnessing the autumn raids of 1940 on Coventry had left a particularly deep impression. As a 'fire watcher' in the clock tower of Birmingham's Council House, Roberts could remember the bombing of the corner of High Street and New Street:

I was up there [in the observation post when there was] one of the worst bombing raids Birmingham ever had and from that position I could see Coventry going up and you could see all the search lights going on over at Coventry and you could see all the anti-aircraft fire and all, and everything going on so that why I was able to say, 'It's going for Coventry tonight, or they, they're coming here sort of thing'. [In] those particular days the Germans used huge . . . phosphorus I think it is, flares, but they, they were big parachutes about oh a good twenty-foot, twenty-foot big parachutes and these things they dropped over the centre of Birmingham which you don't hear about these days . . . they put right up in the sky there and they, the whole of Birmingham was totally white and totally exposed, not a shadow in the whole place, the whole of the city was lit up.

(James Roberts, interview)

In many ways, these recollections chime with recent accounts, which detail the lasting psychological and physical impact of the German air raids on Coventry and which continue to affect the lives of residents, visitors and attitudes of decision-makers (McGrory, 2015).

Consuming reconstruction

Much of the post-war built form of both Coventry and Birmingham continues to be criticised in public, professional and academic representations. Several accounts interpret these cities as being synonymous with well-intentioned, but ultimately 'failed', attempts to implement ambitious, futuristic reconstruction aspirations, emphasising, among other things, cleanliness, speed and order, by effacing many of the ambiguities and threats of the pre-war city (Adams, 2011; Hubbard *et al.*, 2003, 2004). Indeed, for some older residents, recollections often reflected a yearning for the 'better' days before the disruption of the Second World War; a sense of loss also infused these accounts because of the planners' sweeping and rather unsympathetic attempts to 'cure' problems of urban congestion by modernising the overcrowded, insanitary city cores:

Growing up in the '30s . . . Coventry used to have twenty-three cinemas that we all used to go to . . . We used to visit the Geisha café at the bottom of Hertford Street; this was where everyone went. . . . It's only in recent years [that] I've come to think that 'why didn't they keep the history of Coventry intact instead of making these boxes after the war'.

(Phyllis, born 1919, interview)

In Coventry, the task that the City Architect's Department had set itself before the onset of hostilities was to construct a new city centre. This was to include new shopping facilities, services and housing distanced from the perceived constraints of the past and in a much more open, landscaped setting. Achieving this was suddenly rendered much easier by the nature and extent of the bomb damage. The City Council therefore seized the opportunity to plan comprehensively, including a significant increase in shopping provision that had been previously constrained because of the competition with its neighbours, notably Birmingham.

Experiencing the replanning

In October 1945, the *Coventry of the Future* exhibition (Figure 4.1), together with its accompanying (and lavishly produced) pamphlet, provided both plans and perspective views of the new city. The reader was also supplied with detailed information about the types of architectural treatment to be used in the design of the city centre (Larkham and Lilley, 2012). This level of detail was unusual, probably resulting from the drive from the City Architect's department and its influence by European modernism.

It is instructive to consider the number and types of issues raised by visitors to the exhibition: comments recorded in a visitors' comments book had little to do with the overall ambition of the exhibition and were more concerned with mundane matters of 'getting on' and 'getting by'. And despite attracting nearly 50,000 visitors, only 14 letters about the exhibition made it into the *Coventry Evening Telegraph* by the end of October 1945 (see Larkham and Lilley, 2012). Lilley (2004) also notes the muted reaction amongst those respondents spoken to in 2001 towards the proposals for the future city. For example, they reported that, at the time, people felt that more urgent priorities took precedence during the immediate post-war years. Nevertheless, certain respondents suggested that the pioneering experimental 'urban

Figure 4.1 Coventry of the Future Exhibition, 1945

Source: Image copyright of Coventry Archive

utopia' presented in the 'Coventry of the Future' exhibition provided a sense of hope and opportunity for the city:

> My friend and I went to the Coventry for the Future Exhibition at the Drill Hall and that showed the ring road and what it was going to be like. Now I don't know how far they went away from that, but we were quite fascinated by this, this was about 1945 or 46 and about ten years before anything actually started building anything around there, because they started building the Cathedral about 1956, laying the foundations and whatnot. . . . It showed the ring road, and raised ring road, all where the Cathedral was and the rest of it. And we thought it was fascinating. We were all into spaceships, you know . . . Buck Rogers. We thought 'ooh', so we wandered in to here and it was very interesting . . . we'd thought we'd be old men by the time it would get done. We were actually young men by the time it got finished. Aeons of time in the future but it was about ten years.
>
> (Basil, born 1938, interviewed 2001)

> It was models, we went to have a look, you were sort of looking down, you know, like an aerial view. You couldn't visualise it really. Because you had got the precincts, that was the main thing the precincts. And you tried to relate anything with regards to the Cathedral; sort of tried to relate everything, where it was, from that, you see. . . . It was the precinct first and I say we had got all these temporary, prefabricated shops all on the one side of Broadgate and then – of course it was Broadgate she came to – we had the statue of Godiva was in the middle of Broadgate.
>
> (Doreen, born 1921, interviewed 2001)

This view demonstrates how limited visitor perceptions could be in terms of unfamiliar perspectives, spatial location and landmarks – with relatively little of the reconstruction, other than the precinct and the ring road, being specifically referred to. The go-alongs reinforce and extend the narratives elicited from Coventry respondents in 2001. Although none of the go-along respondents could specifically remember the 'Coventry of the Future' exhibition, Raymond spoke of how he remembered a discussion with his father about the wartime damage. Raymond suggested that the bombing had presented an exciting opportunity to build a sanitary, brighter and more orderly 'modern' city:

> I think in actual fact, we were very excited after the war and it was pretty well the first shopping precinct and people from all of the world came to study it, like Plymouth and . . . but this is all new build . . . as my Dad, an old Coventrian who loved Coventry, said that all these people who said that Coventry should have been built in the style it was is rubbish, because most of that probably was rubbish – most of that

probably was knackered anyway, apart from the terrible loss of life. Hitler did a us a favour in getting rid of it, but if you look at towns in France, like St Malo which were built on radial principle it was worthwhile building it because it worked, but Coventry was a real hotchpotch and it wouldn't have been suitable.[1]

(Raymond, born 1937, go-along)

There were similar messages in Birmingham. As Paul succinctly suggested, 'we knew that something needed to be done' (Paul, born, 1946, interview) with what local people perceived to be 'outdated' elements of the postwar Birmingham. Even so, some of the interviewees that were spoken to in 2007–2008 talked rather favourably of the articles and impressions of the future city centre and of the ideas behind the 'New Towns' on the outer edges of the city core:

The *Birmingham Dispatch* that used to be on the corner of Corporation Street . . . [I can remember that] they did a series of artist impressions with what they were going to do with the Bull Ring and other areas . . . [I] think . . . people like my Mum and Dad were looking forward to it . . . [B]y the '60s there was no rationing, and people were wealthier and [Harold] Macmillan said, 'you'd never had it so good' [*sic*] and it was true in a way.

(Steven born 1949, interviewed 2008)

Others recollected that the emergence of the need for a 'new' city; some respondents explicitly welcomed the prospect of having modern housing, schools and other public facilities. This was especially the case among those who could remember (or perhaps lived in) the condition of the slum areas during the post-war years. Steven, for example, who used to travel with his parents into Birmingham from Bromsgrove during the early 1950s, recalled:

[I]t [Birmingham city centre] was all very tatty . . . if you were coming in to Birmingham from the south; if you were coming up the A38 from Bristol or somewhere like that, it wasn't a particularly salubrious area to see . . . [people] would have the niceness, and come over the Lickey Hills and through Northfield . . . [but] it would get progressively worse as got through Selly Oak and through to Bristol Street, it got worse and worse as you came into town.

(Steven, born 1949, interviewed 2008)

This was a view that was felt by Sheila, too:

[Even parts of the city centre were] very dirty and very 1930s because there were a lot of factories and things around there because I was at the Post Office there and it wasn't a particularly nice area.

(Sheila born 1949, interviewed 2008)

Similar messages came through during the go-alongs. For Wendy, the city centre represented a 'curious mix' of architectural styles and buildings; this gave her an overall impression that the city was in 'need of some help' after the war:

> There was Bingley Hall [located on the site of the current International Convention Centre, Centenary Square, off Broad Street] . . . [that was] quite an ugly thing and that was all developed afterwards. [I can remember] the [nineteenth-century] Council House . . . [but] it certainly needed some help, Birmingham did, you know, but we got fond of certain places.
>
> (Wendy, born 1938, go-along)

Experiencing the rebuilding

The rebuilding of Coventry's city centre proceeded at a gradual pace, but despite the reluctance and/or resistance of the Town Clerk (Frederick Smith) and the Treasurer (Sydney Larkin)[2] to underwrite Gibson's proposals, nevertheless the Council persisted in pushing the replanning ideas through. Gibson, ever alert to the possibility of promoting the ideas of reconstruction, ensured the early erection of enduring, visibly striking symbols that embodied the hallmarks of this bold scheme (Gould and Gould, 2016).

These included the levelling stone (Figure 4.2), a 'Savings for Reconstruction' display in 1948, provided the funds to produce and site a high

Figure 4.2 Coventry's Levelling Stone
Source: Authors' own collection

slender aluminium pole mounted with the Elephant and Castle of the City Arms at the top of the Upper Precinct (*Coventry Evening Telegraph*, 22 Apr, 1948) (Figure 4.3). And to accompany Princess Elizabeth's commemorative inaugural official opening of Broadgate, and in May 1948, the lower section of the first column of the Upper Precinct was erected (see Tiratsoo, 1990).

The attraction of Broadgate as a locus for a meeting space – a traditional, albeit congested, market-place and a remnant of the city's medieval past – was inimical to Gibson and his ideas regarding rebuilding. He wrote scathingly of the danger of motorised traffic, the jumble of industrial workshops and factories, variations in materials and architectural styles and the scale/incongruity of interwar premises located at Broadgate. From an architectural point of view, Gibson particularly lamented the 'varying heights, hideous lettering, extravagant squiggles [and] narrow pavements' of the pre-war city (Gibson, cited in *Architectural and Building News*, 21 Mar, 1941, page 192). In an apparent attempt to counter the excess and chaos of the industrial city, Gibson quickly achieved architectural consistency for the proposed more orderly, central shopping area by laying down the scale, storey-heights and construction materials required. This was a new departure for a British city, in which he was followed most noticeably by Patrick Abercrombie's Plymouth (Gould, 2000).

Figure 4.3 Broadgate in *c.* 1980
Source: Reproduced with permission of Rob Orland and Brian Rowstron

Most respondents suggested that they had not objected to the principles of rebuilding but, reflecting on their experiences, were perhaps more ambivalent towards the *process* of redevelopment. It is unsurprising, perhaps, to note that there was a clear sense amongst several people that both cities appeared to be subjected to years of upheaval and disruption. One notable contemporary town planning practitioner with extensive Ministry experience, William Holford,[3] suggested in 1958 that there was little end in sight to Coventry's rebuilding, stating that 'behind the first stage of rebuilding, there is another stage of developing and visible prospects of further stages beyond' (Holford, 1958, page 481) (Figure 4.4).

A perceived 'loss' of heritage and existing ways of life in the city was also apparent in the narratives coming from some Coventry respondents who

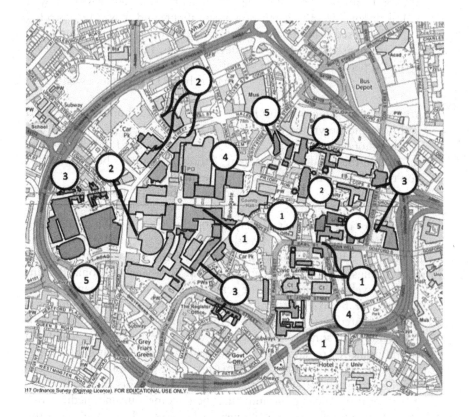

Figure 4.4 Extent of the rebuilding in Coventry. One denotes buildings constructed between 1939 and 1955; two denotes buildings constructed between 1955 and 1964; three denotes buildings constructed between 1964 and 1974; four denotes buildings constructed between 1973 and 1994; five denotes buildings built since 1994

Source: Crown copyright reserved

contributed to the research carried out by Hubbard *et al.* (2003a, 2003b, 2004). There was a certain amount of dissatisfaction towards the apparent lack of sympathy for the pre-war city on the part of the plan-making professionals and decision-making politicians:[4]

> In '46 and at the end of Smithford Street as I remember it and the beginning of Smithford Way as it's called now is with all its later ramps and bridges – I think it's an absolute disaster. But then I'm a traditionalist you see. I've been to towns like Chester and they've preserved it. I mean Coventry originally was described as one of the finest medieval cities in Europe pre-bombing and there was a good deal left of it after the war. . . . I remember the remains of palace yard, I think it was called, just up Much Park Street, which was one of the finest examples of extant timber-framed building in Coventry, and then it was allowed to fall down. Literally it fell down one night, and it was unprotected it just collapsed. I didn't agree with the new concept – I'm ashamed to say that I'm ashamed of the new developments. I think they've ruined a perfectly, perfectly restorable beautiful city.
>
> (Mike, born 1938, interviewed 2001)

> You see when we moved up here, there were [old] buildings, we lived with a friend down the road and they had got these [traditional] bay window houses . . . [the precinct] was very plain, but I did like Godiva situated in the middle of Broadgate. I thought that was beautiful. But I don't like it now. But I did like it [at the time].
>
> (Barbara, born 1934, interviewed 2001)

Gibson was, for the most part, largely opposed to the retention of surviving historic buildings where these directly affected the implementation of the planning ideas. His initial proposals expressed the desire to sweep away all the buildings on High Street including the (pre-war) National Provincial Bank as well as arguing for the removal of the badly damaged Ford's Hospital to the Bond's Hospital site adjacent to Spon Street.[5] Only the Cathedral, situated beyond the main plan area(s), was earmarked for retention. And although some remnants of the city wall, which were scheduled ancient monuments, remained in situ, the radical reconstruction plans suggested the removal of some of these remains. These proposals drew particular criticism from senior staff of the Ancient Monuments Branch and the Chief Inspector of Ancient Monuments, the uncompromising mediaevalist B.H. St J. O'Neill, who made a site visit in December 1952 (see Larkham, 2014b, page 15).

The selective dismantling of the earlier built form was, in some ways, entirely understandable: when plans were first developed they were radical and extensive (cf Hasegawa, 1999) and when actual reconstruction was at its height, ideas regarding conservation/preservation were a decade away from wide public or professional appreciation and application (in the 1967 Civic Amenities Act). Also, from the poor urban conditions of the pre-war period and partially successful efforts at slum clearance, to the increasing use of modernist design principles, it is not altogether surprising why less consideration – although not *no* consideration – was given to 'heritage' in some of the bold ideas regarding reconstruction (Pendlebury, 2003). In Plymouth, for example, 'boldness' was evident in the planners' attitudes to surviving buildings. Aside from four churches and the ruin of Charles Church, only two structures were to be preserved on their original sites. Others, such as a bank building, were recommended to be relocated elsewhere 'to foundations on new sites' as 'some of the important buildings will interfere with the proposals' (Forshaw and Abercrombie, 1943, page 77).

In a comparable way to Coventry, official attempts were also made to manage, control and re-order the post-war urban form of Birmingham. It is tempting to argue that the elimination of apparently unwanted land uses and activities can be taken as an effort to limit the 'strangeness' and dangers of urban encounters (Sennett, 1970 [1990]). While some of the grander ideas were never realised, building activity within the city centre, from the early 1950s until the oil crisis of 1973, was considerable and speedy (see Figure 4.5). Elsewhere in the city centre, there was a continued push towards the provision of more commercial and retail space: Colmore Circus, for example, was the recipient of a group of offices, of which the most significant was the Birmingham Post and Mail Building (John H. D. Madin and Partners, 1960–1965), replete with a 16-storey tower projecting from a podium. There was another shopping precinct above the (rebuilt) New Street station in 1970, which (and furthermore, the redevelopment of New Street station itself) acted as a physical extension of the Bull Ring Shopping Centre. Other 'precinct' developments included the Colonnade development, designed by Frederick Gibberd and widely praised and an unrealised proposal for Queen's Corner (New Street/Corporation Street) in which the influential architect Walter Gropius had some input (*The Guardian*, 20 July, 1962).

Although the city had promoted a private Act of Parliament in 1946 covering Birmingham's inner ring road among other things, there were severe problems with securing central government funding for construction materials (especially steel). Birmingham's first two replacement office blocks in the city centre did not appear until 1950–1953. Elsewhere, however, progress on post-war reconstruction was more hopeful: in a study of nine badly bombed British cities (unfortunately excluding Birmingham), development by mid-1954 in city centres was acknowledged by Rigby-Childs (1954) as being encouraging and by 1955, in the Birmingham conurbation, about 78,000 new homes had been built since the war (Myles Wright, 1955). The inner ring road construction did not proceed until 1957. Treasury funding for the ring road was obtained only after the Corporation put up a stiff fight for what they rightly considered to be a 'vitally important road' (*Architect and Building News*, 1959, page 472).

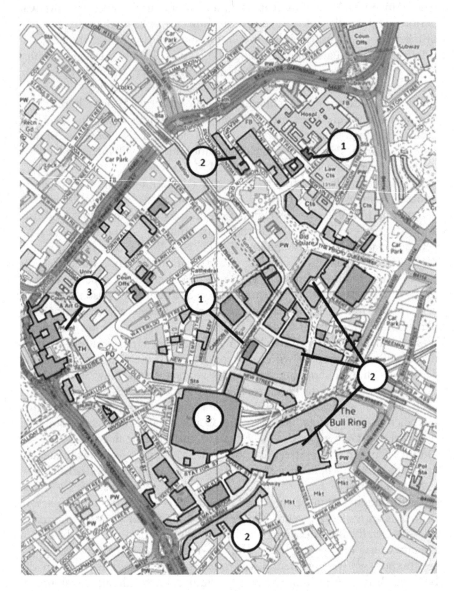

Figure 4.5 Extent of rebuilding in Birmingham. One denotes buildings constructed between 1945 and 1951; two denotes buildings constructed between 1952 and 1967; three denotes buildings constructed between 1968 and 1974

Source: Crown copyright reserved

The phasing and the rapidity of the rebuilding of the city centre during this period were of deep concern to some interviewees: especially those who could remember with some fondness particular buildings that held specific memories. The details of building activity in post-war Birmingham has been chronicled elsewhere (Foster, 2005), but, as detailed below, several respondents had mixed feelings towards the *rate* of change witnessed in the city.

In terms of the architectural and urban form, the *lack* of an overall coordinated city centre plan was felt to be problematic by several respondents, particularly regarding the effects of the redevelopment on the city's townscape. Again, age seems to have been a factor in what people felt about the changes that were occurring as the city began to be reshaped. As much of this activity in the heart of Birmingham's shopping area took place perhaps a decade later than in the centre of Coventry, more of the respondents had strong recollections of the period. For most people this represented a time when there were still remnants of the city's Victorian legacy – and several people viewed the built form from this era with a level of affection. For some respondents, therefore, the newly erected, and overbearingly large, buildings disrupted their 'mental map' of the city:

> All the change that took place, so quickly – [points] that Needless Alley was a joy, all little shops, erm, yes, very old, 'Victoriana', all haberdashery, poky little shops, specialist music shop that you hardly dare to go in, that you daren't go in but they brought it all down and they put all these bigger and much taller modern buildings in [laughs, points to corner of New Street railway station].
>
> (Kathleen, born August 1950, go-along)

For Doreen, the construction of the new Bull Ring Shopping Centre, which replaced the old (and bomb-damaged) Market Hall and radically altered the existing street layout, evoked particularly strong feelings:

> The Bull Ring was horrible, though! We used to sit in Manzoni Gardens to have our lunch – it was nice, but it was a narrow-cobbled street and I know that the market hall everyone called it the old market hall and going down the narrow road that led to the [St Martin's] church, there were stalls, carts, you know. As a young girl and I remember them pulling the buildings down! The subways weren't very nice, though. . . . When they altered this, down steps under a horrible tunnel and then a slope – when my sister worked in Waterloo Street, we would go and have their lunch in the churchyard if you were in that part of town, but Manzoni Gardens were here, and they moved Nelson's monument.
>
> (Doreen S., born 1945, go-along)

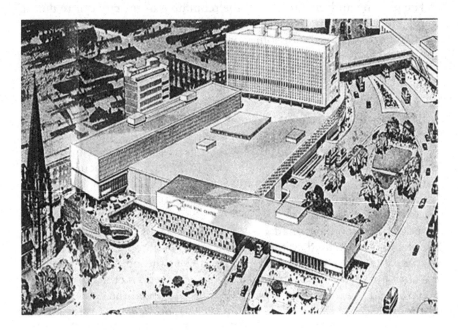

Figure 4.6 Artist's impression of the Bull Ring Shopping Centre
Source: Laing Development Company, 1964

Tom, in his 20s at the time of this rebuilding, recalled his displeasure with the alteration of the historic Bull Ring area (Figure 4.6):

> Coming up to the sixties they kept some of the old character but then they put new in and they changed it all completely. If they had kept it with the old lamps, street lamps that would've been nice kept the character like. But they took all the old buildings out and the character was taken out of it.
>
> (Tom, born 1942, interviewed 2007)

Iris also reported strong feelings of loss: for her, the new Bull Ring Shopping Centre 'spoilt people's lives':

> The biggest upheaval was the Bull Ring Centre – it was absolutely massive – it spoilt our lives really. It was for the Birmingham people, all the markets were there, the meat market, the veg market and there were lemons, oranges, the flower lady and they bombed the market hall and. . . . That used to be our play ground – we were living in Broad Street, born in Cregoe Street and we grew up with town people and we used to meet the same people to go socialising, meet in Lyons

tea shop and it was just Nelson's column and it was just a cobbled street, really.

(Iris, born 1934, go-along)

A sense of enthusiasm

Several Coventry respondents, who contributed to the round of oral history interviews conducted in 2001 (see Hubbard *et al.*, 2003a, 2003b, 2004) – especially those who were younger at the time of the post-war rebuilding – also felt that there was a palpable sense of excitement about shopping in a largely traffic-free pedestrian precinct area. For these people, the new city centre, especially in the 1950s and 1960s, provided an accessible, salubrious, verdant, safe and convenient shopping experience and a sense of 'freedom to wander around'. For some, therefore, the new centre achieved the aspirations of its designers. By the late 1950s, the city's booming manufacturing and engineering industries were prominent wealth generators; the population was also young and prosperous:

> I thought it was great, it was quite exciting all these lovely new shops, open spaces with trees, you know, we'd never seen anything like this before. . . . And also, no cars, you didn't see the cars, so you were able to walk freely about and of course later on when I got my own children, it was always easy to take the children in town because you never used to have to worry about the cars. But it was so modern and so nice for its time, . . . but it was pleasant, there were places you could sit, there was the trees, there was the flowers, there was the shops and like I say there was no traffic, it was great.
>
> (Jean, born 1936, interviewed 2001)

> There was no cars. [But before the war] there was no walking in a town or city with no cars or bicycles or anybody driving around in the centre of town. Everything was as I say all concentrated – as far as movement was concerned – was up in Broadgate. Nowhere else. So that was quite a thing I felt was wonderful really that the way the city was being put together and being built by having just pedestrian areas. It was a city without any cars and bicycles in the middle of town. So, you could walk around, quite enjoyable. . . . I think that that was something quite impressive in those days.
>
> (Gerald, born 1936, interviewed 2001)

The sense of excitement was further underscored during the go-alongs. Many respondents were in their teens or early 20s at the time of reconstruction: one respondent suggested that there was a very tangible public mood of enthusiasm about Gibson's ambitions. Hugh pointed out that the plans for rebuilding had a very profound 'sociological impact' on Coventry. For him, the process of rebuilding had a positive impact for several companies intimately bound-up with reshaping the city's post-war future:

> I think people were interested in what Gibson was doing and I think it was a huge sociological change in Coventry brought about by the war – the damage – and lots of people, builders from all over the country, Ireland, traditional trades came in and set up successful building businesses off the back of the redevelopment and then subcontractors – O' Flanagan's – and then built it up the very successful house-building business as a result. This had a really deep impact and then the other thing alongside the Cathedral. . . . But yes, people quite liked initially the Gibson scheme and it was quite revolutionary.
>
> (Hugh, born 1941, go-along)

Whereas 1950s Coventry had a relatively high proportion of 'men' employed in vehicle and engineering sectors (67% in June 1956), it is arguable that Birmingham's industrial base was more diverse at the start of the decade. Although 23% of the total working population were employed in engineering and vehicle manufacture, 11% were engaged in the distributive services industry and 11% in the insurance, banking and finance and other professional services sector (Coventry City Council, 1958). This perhaps goes some way to explain why there was certainly a continued push towards the provision of more commercial and retail development in the city during the following decade was. Birmingham residents recorded similar experiences. For example, Peter, who was born in 1930, recalled that there was a feeling of enthusiasm towards the extent of the rebuilding during the 1960s:

> [It had gone] from small Victorian and bombed buildings if you like, to suddenly these vast building projects. Not only these buildings but also the inner ring road because this came about in chunks, and with each section . . . you know, brilliant! You thought, 'what has happened here?!' . . . there was these great façades of glass and concrete.
>
> (Peter, born 1930, interviewed 2008)

Those 'great facades' were clearly evident – bright, clean and white – on aerial photographs of the city centre from the 1960s and 1970s. Steven felt a similar level of excitement at the changes being made to the city centre:

> The whole of town was under construction [in the 1960s]. When I worked in Lewis's [Corporation Street] – you were generally walking a different route every day because you had to avoid the construction works, over plants, and if you went to New Street Station, it was a different way in and a different way out every day (laughs)! But we saw that as being quite exciting really, and the sorts of facilities that were being thrown up, but we hadn't reckoned on the Queen's Hotel being

pulled down (laughs). I think that [overall] people [of our generation] were rather pleased with it and that it was a world-class facility.

(Steven, born 1947, go-along)

Design, layout and function

The design of the rebuilding of both Coventry and Birmingham was, to a large extent, a local affair. Coventry's City Architect's Department designed and carried out (or led) the development itself. In terms of the actors involved with shaping post-war Coventry, established architects from outside the city were largely ignored: the City Architect's department designed notable municipal buildings, together with large swaths of the shopping area, in-house. Developers also drew on the experience of local architects such as W. S. Hattrell & Partners and Redgrave & Clarke. There was some thoughtful consideration given to this idea: Gibson argued that the overall integrity and consistency of the post-war plan must be maintained, thus avoiding a confusing assemblage of different building heights (such as, he argued, had existed before the war) and different materials all facing onto public spaces and streets (Campbell, 2004; Pickford and Pevsner, 2016).

Gibson did follow Le Corbusier's 'four principles' identified in his polemic, *City of Tomorrow* (1929), as a means of guiding the post-war planning process; however, the notion of building high to relive the density and congested nature of the urban core was deemed to be inappropriate for Coventry. As Johnson-Marshall (1966, page 293) argued, new buildings were purposefully 'kept low in order to emphasise the verticality of the Cathedral' and, by implication, the other two spires for which the pre-war city skyline was known. There is some commonality here with the ideas developed by Thomas Sharp for Exeter when he spoke of the importance of 'opening up monumental vistas [of the Cathedral]' (Sharp, 1946, page 90). Similarly, at Salisbury, Sharp argued for one 'direct view [of the Cathedral] . . . informal and in character with the rest of the city' (Sharp, 1949, page 36). Although the Architect's Department had a degree of influence in promoting the widespread use of Blockley Brick interspersed with Westmorland slate and Hornton Stone, thus giving an overall appearance of uniformity (see Figure 4.7), Trevor, for example, offered a rather withering attack on the overall nature of the 'square', 'blocky' and 'uninteresting' design of the new buildings:

> When they were first built, apart from the square business, and I thought well, surely they should have put some curves here and there and maybe the odd pillar or two with a curve. I think the eye is more attracted to a curve than a square. It was modern and replaced a lot of junk. . . . So, the development took a lot away from that and brought airy buildings so when you're inside one it's quite pleasant, it's quite airy, it's well-lit.

> But, outside, the structure to me, [was] too square, much, much too square. And I'm not very much in favour of it.
>
> (Trevor, born 1933, interviewed 2001)

Despite Coventry City Architect's Department setting the general design principles for the 'new' city centre, the private sector and non-local authority architects were also responsible for developing 'square' designs of certain of the prominent post-war buildings. These designs would have to conform to overall principles set out by the city council, yet local firms of architects were responsible for the design of the Hotel Leofric (W.S. Hattrell & Partners) and the Owen Owen department store (Rolf Hellberg and Maurice Harris) (*Architects' Journal*, 8 Oct, 1953) (Figure 4.7). The 'link blocks' (North and South blocks) of the Upper Precinct (started in 1954), were designed by W.S. Hattrell & Partners, employed by Ravenseft Properties Ltd., a recognised pioneer of the post-war reconstruction of Britain's shopping centres. Ravenseft Properties Ltd also was involved in rebuilding blitzed cities such as Birmingham, Hull, Exeter, Plymouth, Bristol and Coventry and carrying out many developments in close association with the New Towns (e.g. Harlow, East Kilbride, Basildon) (see Marriott, 1967). Criticisms were levelled at the unsympathetic architectural design of the new shopping precincts.

Figure 4.7 Owen Owen store
Source: Authors' own collection

As with the 'blandness and monotony of "Stalinist" architecture' (Punter, 1990, page 30) associated with the post-war rebuilding of Bristol's Broadmead shopping centre, certain respondents felt:

> [I]t was lots of modern shops with lots of glass frontage which was very un-Coventry. Old Coventry shops were Edwardian type ornate curved glass frontage then great vast expanses of plate glass, featureless. And people didn't like it. It was, you see, the old Coventrian people who had lived there before the war from the '20s. They had been born in Coventry and had lived through the bad years of the '20s and '30s and they were not totally impressed with what was happening around them.
>
> (Mike, born 1938, interviewed 2001)

For those people who were younger at the time of reconstruction, Coventry's precincts also represented a place of adventure, though this feeling was perhaps tempered by a feeling that the shopping area was initially 'alien'. This perspective was particularly true for Valerie:

> [I can remember] the noise of people . . . by the 1958 most of the precinct was built and . . . what I never like it, it just didn't feel right [to begin with], but it's like everything else in life, you learn to accept it. It just didn't feel right to me. [Although] it was a bit of an adventure because young and lively you obviously went up to the other levels from BHS and from Marks & Spencer and you could once you were on the level you could you know go to both of those places quite easily.
>
> (Valerie G., born 1931, go-along)

Contemporary critics of the architectural design of the Gibson-era buildings were also rather less scathing in their judgement. Rigby-Childs and Boyne (1953, page 430) considered that the use of 'native materials . . . and of interesting, almost picturesque shapes and massing gives a "friendly" looks to the building group'. In the *Architectural Review*, J.M. Richards (1952) openly conceded that the Broadgate was one of most impressive piece of civic planning that has been achieved in British post-war planning. But notable attempts were made by Coventrians to draw attention to the more destructive aspects of post-war reconstruction. The formation of the Association of Coventrians in 1953 represented an attempt by residents to combat the 'lack of consciousness' displayed by the 'planners', with founding member, Abe Jephcott, arguing that the 'atmosphere' of the 'ancient city' of Coventry should be brought back (*Coventry Evening Telegraph*, 1941, page 6).

Others were perhaps less convinced about the design of the precincts. One female respondent, who could remember the pre-war city, explained that locating the shops in an essentially mono-functional area presented a challenge to service vehicles trying to negotiate their way through to deliver goods:

> We liked the traffic free business, I mean, that was a big point, you know, all no traffic away from the road, but then they started bringing delivery vans in. They've stopped that now, but they used to say that they couldn't get around the back to deliver and all the rest of it. So that took the gilt off the gingerbread.
>
> (Doreen G., born 1921, interviewed 2001)

Regarding this specific point it is worth remembering that the Chamber of Commerce plan proposed entirely separate servicing arrangements underground – although it was suggested that groundwater levels would make this impractical.

Though Gibson both created and pursued the early realisation of the plan, he left Coventry in 1955 'increasingly frustrated by obstructions to Coventry's architectural progress' (Saint, 2004, page 67). The later City Architects, Arthur Ling (from 1955 to 1964) and Terence Gregory (1964 to 1973) added their own individual interpretations to the original plan. Ling, for instance, pedestrianised Market Way and Smithford Way, while he also recognised that the two-level shops of the Upper Precinct had caused some inconvenience among shoppers, and hence, it resulted in a loss of trade. Residents spoke fondly of Ling's attempt to enliven the precincts with the creation of two pubs and the Locarno Ballroom. There were also pockets of enthusiasm for Gregory's 'sharp and contemporary' Bull Yard (Gould and Gould, 2016, page 72). Nevertheless, there were some public criticisms of the later 'block-like' buildings and, in particular, Ling's bolder architectural approaches:

> The land-use zoning did exclude the human factor, perhaps, and traditionally, people would have lived above shops, but the tower blocks at the end of the precincts felt more like an afterthought – the one in the lower precinct came later.
>
> (John, born 1931, interview, mentioning Mercia House, built in 1968 and designed by Arthur Ling and Terence Gregory, City Architect with North & Partners of Maidenhead)

Birmingham's planned and actual developments had a very mixed press, for professional and lay readers, in national circles and locally. By the onset of the oil crisis in 1973, there was very tangible public anxiety reacting against the scale and design of some of Birmingham's post-war buildings. For instance, the architectural critic Ian Nairn supported some of the civic centre proposals; however, he also conceded that the 'rebuilding is imposing at a quick look . . . disastrously ham-handed as it appears in the concrete'

(Nairn, 1960, page 115). There were concerns in over the proposed widening of Colmore Row incorporating a section of St Phillip's churchyard; this would have involved the demolition of several existing Victorian buildings on the road's southern edge (Larkham, 2007). Criticism of the drive towards modernisation and car-centric planning ideas was felt elsewhere by the mid-to late 1960s. James Stevens Curl (author of *The Erosion of Oxford*) in a piece published in the *Oxford Mail* suggested that 'those of us who cared for the architectural values of the past were very much in a minority, while roads, traffic and non-traditional architecture were all accepted as essential aspects of "progressive" thought' (Curl, 1968, viii). The reaction of Councillor D.S. Thomas, Chairman of the Public Works Committee, typified Birmingham's response to post-war reconstruction: 'Birmingham had at best avoided the mistake of failing to achieve anything at all through trying to plan paradise on a piece of paper' (*Architects' Journal*, 15 Oct, 1959, page 547).

A former Lord Mayor of Birmingham, Norman Tiptaft wrote in 1960 that 'there is a deep, underlying resentment at the way the [rebuilding] is being done . . . to accept that these changes have met with the complete approval of the population, with the possible exception of a few nostalgic old fogeys. That is not so'. He suggested that designs were 'hardly designed at all' and that both commercial and housing developments were overpriced: arguing that 'there is a feeling that citizens have been handed over to developers more concerned with financial profits than with building a city beautiful' (Tiptaft, cited in the *Daily Telegraph*, 23 Nov, 1960). A few years later the then Lord Mayor, Alderman Frank Price, said that the city had been 'thrust . . . further forward in redevelopment than any other city in Great Britain' (*Birmingham Evening Mail* 8 Apr, 1965). But a local councillor, Anthony Beaumont-Dark (later a local Conservative MP) suggested that the newly opened Bull Ring Centre was 'the biggest white elephant in the history of Birmingham' (*Architect and Building News*, 1965, page 1074). Owen Luder (later PRIBA) likewise felt that the centre was uninspiring, poorly related to the rest of the town centre and unduly dominated by a complex road plan. The basic lesson from the Bull Ring should be 'that when roads and buildings are integrated in a three-dimensional layout they must be designed at the same time. To try to make the buildings fit afterwards as in this scheme is hardly likely to be very successful' (Luder, 1964, page 401).

Of our respondents, Donald, for example, was particularly critical of the way in which the city's nineteenth century legacy had been unceremoniously/unsympathetically replaced:

I think one of the criticisms that we do have about Birmingham in the past that we used to have a lot of old buildings like the Post Office in Victoria Square, which given half a chance the planners would have demolished and a lot of fighting to keep that building. And I think a lot of other buildings could have been kept – we lost them, they just wanted to get rid of them. Possibly nowadays they would look a little differently about it and try and preserve them.

(Donald, born 1942, interviewed 2008)

In fact, Foster (2005), for example, suggests that the early 1970s public campaign against demolition of the Victorian Post Office (Figure 4.8) helped to harden attitudes towards the scale and pace of urban development.

As most of the Birmingham go-alongs started from the Council House located in Victoria Square, respondents were keen to point out their feelings towards the city's Central Library (demolished 2015–2016). This building was a short walk away situated in Chamberlain Square. While standing outside the library, several respondents lamented the removal of the previous Victorian central library (built in *c.*1882) which was demolished in the early 1970s to make way for the John Madin's Central Library. Iris, for example, who grew up on the nearby Broad Street, recalled with some affection the 'old' Victorian Central Library:

It's really hard [to remember] but the old library used to be behind the library there [points] and it was one of the nicest buildings you would ever wish to see! Lovely inside. . . . You can hardly tell where everything

Figure 4.8 Post Office, Victoria Square
Source: Authors' own collection

was now. It was one place where I was allowed to come on my own. It was beautiful, very quiet, and very majestic. There were loads and loads of books as you can imagine, everyone would whisper, and you didn't dare speak out. But it's these dreadful new buildings [points to the Central Library] and it used to be so lovely, fountains, I mean look at that there it is just so awful! I mean look at it, it is a monstrosity!

(Iris, born 1934, go-along)

The library's Atrium was covered in glass and shops constructed in the late 1980. However, for Kathleen, the Madin library retained a certain 'cold' and 'unfinished' quality. After touching one of the concrete uprights (Figure 4.9) she suggested that:

Actually, my memory of this space [atrium of Central Library] was that it was quite bleak. . . . This stone [touches concrete], and they put these buildings in with shops and I think there was a cafe here (left hand side) and an ice-cream parlour – there were lots of window boxes, is the best way to describe them, all coming down. But it was all concrete; it seemed cold and unfinished, with all the pipes going through [points to the piping].

(Kathleen, born August 1950, go-along)

In fact, several respondents of the Birmingham go-alongs were scathing of the visual appearance of the Central Library. For example, Valerie was less

Figure 4.9 Madin's Central Library
Source: Authors' own collection

than sanguine about the design of the library – especially as it represented something that was so visually dissimilar to its predecessor:

> You know, people didn't like concrete, it was just concrete, and people just didn't like it, and yet, I actually feel very proud of the fact that we have the biggest library in Europe, in one way at least. The previous library was a beautiful building. The other side of the fountain, the stair case on the sixth floor of the central library was the staircase that was in the original Victorian library, was out of that building. It was beautiful. It was magnificent, really. A lot of people didn't like the demolition of the building. . . . Mind you, they were going to clad in stone, but they ran out of money and it never got done.
>
> (Valerie P., born 1932, go-along)

But there is nuance to this debate. John Madin was asked in 1964 by the then City Architect, Sheridan-Shedden to collaborate on a new civic centre master plan, combining an ensemble of civic buildings, including a new library, at the eastern end of Broad Street on the site known as Paradise Circus (Sutcliffe and Smith, 1974). Madin produced a large model, showing (among other buildings) the Town Hall of 1832–1834 and the Hall of Memory war memorial, together with a bus station, student halls of residence, a concert hall and library. Madin's plans for Paradise Circus were approved by the council in 1968, and the original scheme was for a central library, with a bus terminus underneath, a school of music and physical sports institute – this was Madin's 'civic heart' of the city (Clawley, 2011). Construction of the library began in 1969 and the main shell of the building was completed in 1971. The outward form was simple and comprised a huge reference block and smaller lending block to its east, which also housed the first set of escalators leading to the upper floors of both libraries (Figure 4.10). Adopting a cantilevered design, each floor was larger than the

Figure 4.10 Birmingham's Victorian library (right) and Madin's Central Library looking from Chamberlain Square

Source: Reproduced with permission of the Library of Birmingham; authors' own collection

one below resulting in a distinctive inverted ziggurat formation. This was adapted for civic purposes in the monumental Boston City Hall design by Kallmann, McKinnell and Knowles, in 1962.

Madin's original vision of a building dressed in Portland stone or travertine marble, set in landscaped gardens replete with fountains and waterfalls, was altered, and pre-cast concrete with a stone aggregate offered as an alternative by the City Architect was adopted instead, leading to some subsequent criticism that the library was a 'concrete monstrosity' (e.g. Parker and Long, 2004). The Council also cited the failure of some of the concrete panels in 1999 as a reason to demolish the library and pass the site to a commercial developer (Clawley, 2011).

Although more as a part of the broader improvement and electrification of the West Coast Main Line than as a direct response to the wartime destruction, the bomb-damaged New Street Station was also redeveloped. The main entrance was aligned towards the line of the inner ring road, with all services being at ground level and platforms below ground. Work began in 1964 and the whole £4.5 million station was opened on 6th March 1967 (McKenna, 2005, page 133). Above the station itself a further 7.5-acre concrete raft supported a new shopping centre with 96 units by Cotton, Ballard and Blow, costing a further £6 million and built in 1968–1970 (Larkham, 2007). While walking along Corporation Street towards New Street Station, Doreen pointed out that the demolition of the original station and the subsequent rebuilding was particularly memorable:

> The main change when my ex-husband met on the train and through our courtship they pulled-down the old New Street station and built what is there now [points]; but that's changing again and the big Queen's Hotel and all that next to it had to be demolished to build a huge great shopping centre [above New Street], that was quite sad because that's where we used to meet when we were courting, you know. That was really sad for us [to see that go].
>
> (Doreen S., born 1945, go-along)

Alongside the new architecture of the precincts, the 're-packaging' of Coventry's medieval Spon Street (Figure 4.11), lying between Corporation Street

Figure 4.11 Spon Street
Source: Authors' own collection and sketch

and the western edge of the ring road, generated very mixed feelings amongst some respondents and others about the perceived unsympathetic treatment of the city's heritage generated some concern that the area has become an 'open-air Museum of Quaintness' (Taylor, 2015, page 302), since the Council's decision to move medieval buildings from their original sites (in the late 1960s) and re-erect them elsewhere (Pickford and Pevsner, 2016, page 265; *Architects' Journal*, 1975; Gill, 2004). For Mike, the repositioning of these buildings was problematic:

> I thought these buildings should have been preserved and not moved wholesale to Spon Street where they are totally out of context, where they never stood. That was a watch-making area. It wasn't medieval shops made of timber.
>
> (Mike, born 1938, interviewed 2001)

John A., for example, could affectionately remember the street before these alterations:

> This is Spon Street town plan where several buildings have been put there from elsewhere in the city; where that building just out there was [points]. . . . That is where I served my apprenticeship, but this area was primarily a watchmaking and dyeing area. The company I worked for, they were silk dyers, they introduced these mulberry trees with the idea of producing silk worms and never worked, and now all across there were courts. This has just completely been altered! It was a densely populated area and people used to be a place called Conduit Yard and from all round here people would come to get their water from here and pumped from St Catherine's well a mile and half away and that's one of the town's very old roads that Thomas Telford built – the Holyhead Road.
>
> (John A., born 1937, go-along)

Other respondents touched on the wider social and economic forces that were also important during the post-war reconstruction (see also Johnson-Marshall, 1966; Campbell, 2004, 2006). Furthermore, Rigby-Childs and Boyne argued that the 'new' city centre was necessary to suit the requirements of a 'new' population:

> All this supposition [regarding development] ignores the possibility of a change in outlook of the inhabitants. The bulk of the inhabitants are working-class men, ninety per cent. Of whom were not born in the city. The natural ties of affection for a city, due to being born and bred in it, are therefore absent, and the urge to get the best out of, and into, the city, not yet fully developed.
>
> (1953, page 443)

John H., for example, was also keen to stress that fact that the proportion of people employed in general engineering and vehicle manufacturing was far higher in Coventry during the 1950s (and 1960s) than the national average.[6] This was an important point that should not be ignored when considering the context of reconstruction – it had implications for the makeup of the city's population:

> We always say we a great pride that this was unique and totally car-free precinct rather than adopting existing street patterns. [This is] not entirely true . . . there were cars – cars were accepted, to an extent. . . . What made Coventry different, though [from other cities], was that Coventry was the town that created [and] lived cars and the car owner-ship here was probably the highest in the country. But the basic sala-ries in Coventry was double – astronomical wages and salaries on the back of the car industries and making the unions very strong – increase wages, and this slopped over to the rest of the community.
>
> (John H., born 1931, interviewed 2012)

For Brian Redknap, the decline of the once flourishing car and machine-tool industries by the forces of deindustrialisation during the 1980s resulted in the general decline of the shopping environment, the types of shop on offer and a weakening of public attitude towards the post-war city centre:

> We eventually got a BHS, but the quality of shopping went down and the decline of the motor industry, and the closure of the GEC, as a major employer, [this] reflected in the appearance of the town . . . [Gibson] – really misjudged the regional influence of Leicester, Leamington and Birmingham and extremely less wealthy groups of people. Whether he thought that these proposals would be so imaginative that that would create the wealth, I don't know, [but] by 1970 a good deal, a fair pro-portion of it [had declined] – people looked at it and decided no and there is much more exciting to invest in. . . . There was huge enthusiasm and it had a sort of logic, but it was the decline of the wealth generators and . . . by 1970–75 things that were once fizzling grounded to a halt and started going into decline.
>
> (Brian Redknap, interview)

Speed, efficiency and movement

There were key similarities in how both Coventry and Birmingham approached building inner ring roads. As with other towns and cities charged with wrestling with the issue of traffic management, many existing roads were either widened or straightened, thus opening new possibilities for more people to access the city core (see, for example, Gunn, 2011, 2018). Despite the general acceptance that the inner ring roads were of integral importance for the better functioning of post-war city centres, Escritt critically reviewed the 'craze' of ring road planning and construction. He demonstrated that, with few exceptions, ring roads were not carrying substantial traffic loads (Birmingham was one of the exceptions) and called them 'superfluous and therefore wasteful' and 'irrational' (*The Builder*, 1945, page 519). Nevertheless, both Coventry and Birmingham pressed ahead, but the economic difficulties of the immediate post-war period resulted in significant delay to their inclusion into the detailed construction programming, and serious attempts at implementation did not start in earnest until the late 1950s.

In the mid-twentieth century 'freeway era', when the city became organised around the motorcar (Hall, 2014), municipal governance arguably served to normalise automobility. For those owning cars, the consequence of this politic introduced a distinct form of movement that seemed fast in comparison with the languorous experiences of the pedestrian. From the pedestrian's perspective, this differential mobility has been interpreted by some as disenfranchising: urban cores simply became less welcoming, with walking arguably being viewed by 'planners' as being subordinate to motorised traffic as a means of transport. Moreover, the emphasis placed on creating infrastructure that could ensure free movement of people, goods and capital also has some connection with Robert Moses's high-modernist approach during the mid-twentieth century to the rebuilding of New York, where inner-city 'villages' were cleared to make way for freeways (Berman, 1983; Harvey, 1982).

Birmingham's pre-war street pattern was deemed inadequate for the demands of contemporary traffic, as some of those roads at the historic heart of the city (such as Smallbrook Street) were particularly problematic to facilitate the flow of modern traffic. Drawing on the biological metaphors of the body, Manzoni commented in 1958 that congestion 'started in the organs of this country, which are the urban areas, and, like cancer, it has spread . . . fortunately, unlike cancer, it is curable' (quoted in Merriman, 2008, page 66). Connecting to wider contemporary concerns that traffic congestion was dangerous, noisy, polluting and wasteful in terms of fuel, Manzoni therefore bluntly argued that the design of the inner ring road should help to relieve the choked lines, suggesting that 'motorists should not bring their cars into the city centre' (Manzoni, 1968, page 3).

Most respondents certainly suggested that the city core during the 1960s was a place of danger because of vehicular traffic. This was particularly noticeable at the confluence of Colmore Row, Victoria Square and New Street. It is salutary to point out that during the early 1960s Colmore Row carried five A-roads of traffic. Maggie recalled that walking from her place of work at the Town Hall during her lunch breaks along into New Street was fraught with difficulties:

> It was a square [Victoria Square] and as I say, the Lyons tea shop [on the corner], bank buildings and official buildings in the centre it was all traffic. The road went round and down New Street, down Colmore Row that way and down the square and into New Street: it was incredible, but the city had great problems; it was just all traffic!
>
> (Maggie, born 1937, go-along)

Alongside the desire to ensure a 'continuous flow' of traffic in the city centre, Manzoni also acknowledged that the second consideration was that the inner ring road should 'provide for the extension of the commercial and shopping centre of the City – which is too small – . . . and the intention was also to provide a means for termination of the arterial bus routes, and for turn-round or exchange points on the edge of the enlarged area' (Manzoni, 1961, page 261). Manzoni envisioned that the design of the road was consciously chosen to strike a balance between the 'needs of traffic, shopping and commerce, with due cost and amenity' (Manzoni, 1968, page 269).

Coventry's circulatory traffic system and pedestrian precincts provided a widely adopted 'model' for city-centre redevelopment, and as Rigby-Childs and Boyne (1953) note, the city came to be regarded as a 'test bed' for comprehensive planning and for grappling with the issue of the growth of the private motorcar (Richards, 1952). As with the design of Birmingham's inner ring road, there are obvious connections to be made here between Gibson's conception of how the city could be deliberately compartmentalised into functional and ordered units and Le Corbusier's grandiose vision of engineered streets cutting through the apparent urban disorder enabling the unhindered movement of vehicular traffic. There may have been some link with 'idealised' geometric city representations in the imposed geometry of ring road alignments: both Manzoni and Gibson used idealised circular city plans in their public lectures; and the Coventry Chamber of Commerce's semi-circular plan was at least a semi-serious proposal. Redknap (2004, page 36) points out that the inner ring road, as originally conceived in Gibson's 1945 Plan, was intended to alleviate a growing concern of increasing vehicle ownership and traffic levels in the city. The road was designed to be a one-and-a-quarter mile encircling dual carriageway boulevard, running through 'decaying residential and industrial areas' complete with surface roundabouts at the junctions connecting to internal roads, including links to the inner circulatory road and the main precinct, and green reservations with cycle and footways were originally designed for both sides.

With the Cathedral positioned at the centre, Gibson proposed a system of 'radial and ring roads . . . using existing roads wherever possible . . . with a view to accommodating the increased road traffic of the future . . . noting that it would be fitting for the centre of the motor car industry to give a lead' (Gibson, cited in *Journal of the Royal Institute of British Architects*, 17 Mar, 1941, pages 76–77). As with Birmingham, concern about the rise in motorised transport was evident: 'there are two local factors influencing the growth of traffic in Coventry' Pugh and Percy (1946, page 80) argued: 'the motor mentality' of the city and second the 'future growth of the town' and that in '1938 the number of private motor vehicles per 1,000 inhabitants was 68 – considerably above the 1939 national average of 39'. Brian Redknap keenly stressed the importance of Ernest Ford's pre-war plans to alter elements of Coventry's city centre: these involved the construction of Corporation Street, Trinity Street and the widening of High Street and The Burges. According to Redknap, the destruction of streets and buildings between the Cathedral and Broadgate – the heart of the city, including Butcher Row, Little Butcher Row and the Bull Ring – irrevocably changed the character of the city centre *before* the wartime destruction:

> Trinity Street and Corporation Street were built and there was a huge amount of destruction . . . [There were] huge numbers of timber-framed buildings and what these new streets did was that it introduced additional highway capacity and allowed Owen Owen to take place. Then in 1938, the Coventry bypass was built, but up until then it was medieval street pattern.
>
> (Brian Redknap, interview)

For Redknap, Coventry's inner ring road also offered a particularly innovative solution to the combined pre-war problems of traffic congestion and increasing road casualties. The need for an inner ring road was obvious:

> Like most cities, it had a congested old centre and then there [was] a whole belt of poor quality factories and working-class buildings and this [belt] it more-or-less defined a route and we were already some way from the war and the thought was to push it further out [from the city core]. Certainly, planners were increasingly obsessed about building a ring road at all and it was a continuous background and to some extent the city engineers department too.
>
> (Brian Redknap, interview)

Overall, there were largely positive reactions among most respondents towards the time-saving attributes associated with the inner ring roads. For busy drivers and pedestrians, the inner ring roads and their associated infrastructure helped in this regard: one Coventry resident reported that:

> I mean we did like using the ring road. You know I had got my Lambretta and I could easily get from one side of town to the other very, very, quickly and so basically, you see, as young people we thought it

was worth the sacrifice. Because it made life easier. We didn't want it, but now it's there we use it and it makes life more convenient in many ways, except for the poor people which have had it built straight across the back of their houses and spoilt their way of life.

(Lyn, born 1939, interviewed 2001)

This was the view shared by Philip, though he also hinted at the constrained nature of the road and how the road meant that key public spaces like Lady Herbert's Garden became 'crammed up' against the ring road:

I think basically [the ring road] was and still is a good idea, though it took some time to get built! But you could get to all parts of the city relatively easily and quickly, get in and then get out! But [it is] not large enough and it needed to be further out because it was too close to the centre, [. . .] I always used to park outside [of the city centre], and I had to go under subways. [Of course], you could be mugged under there very easily, and they've put cameras in, but the principle of the subways was all right.

(Philip, born 1938, go-along)

Mixed reactions

Both proposed and developed projects had a very mixed public and professional reaction, though even by the 1960s there was already an emerging criticism of the seemingly diffuse approach to the reconstruction. Ring roads were a particular target: the hugely influential Buchanan Report on *Traffic in towns* (1963) was critical of the fashion for ring roads, although at that time few had been completed. Writing with particular reference to the reconstruction of bombed towns, Buchanan commented on the 'peculiar difficulties' of such roads, with their 'rigid encompassment of the centre, severance of the central area from the rest of the town, frequency of major intersections on a comparatively short length of road, encouragement of heavy traffic along old-type radial roads and the problems of constructing [them] through densely developed property' (Buchanan, 1963, page 210). In Birmingham, it was reported that 'there is no apparent pattern in the redevelopment at the heart of the city' (*Birmingham Evening Mail*, 8-Apr, 1965).

There were several criticisms of the design of Birmingham's inner ring road, and negative professional comments were directed at the design of the Smallbrook phase, with the buildings closely lined with buildings in traditional urban form. Although the overall scheme was 'conceived within an engineering tradition, as an engineer's vision and driven through with enormous energy' (*Drive*, 1971, page. 53), the contemporary commentator Aldous offered a cooler interpretation of the scheme. He suggested that the road was 'essentially an engineer's strategy, first functional and only incidentally as an afterthought concerned with aesthetics or social fabric' (Aldous, 1975, page 26). Frank Price later criticised the level of control exerted by the Ministry of Transport, whose 75% grant was vital for the whole project: 'I was to regret the fact that the Ministry . . . because they held the purse strings, gave cause for a number of fundamental mistakes to be made in the design of the road' (Price, 2002, page 149). In one of the most pointed attacks directed towards the design of the ring road, the architectural critic Leslie Ginsberg, Head of Birmingham School of Planning at the time, wrote that:

> Unhappily this looks like being the greatest traffic and town design tragedy yet to afflict an English city. There does not appear to have been any real traffic survey, or assessment of future probable needs: only the most limited volumetric counts and the feeling that a new pipeline would somehow clear the other choked lines. There is no attempt to keep pedestrians away from the road except by means of ugly underpasses at junctions.
>
> (Ginsberg, 1959, pages 289–290)

Interestingly, Manzoni and Price 'wanted the planned underpasses to be for traffic and not pedestrians, but the Ministry of Transport were adamant on this point too. . . . There was nothing that we could do about it' (Price, 2002, pages 156–157). Any concerns for accommodating existing pedestrian routines – particularly for those moving from areas immediately beyond the city core into the centre – were seemingly overridden by City Council's desire to improve the economic efficiency of the city. This was particularly noticeable, perhaps, during a meeting with the Ministry of Transport in 1956. Council officials strongly argued that money from the Treasury was required to ensure the unconstrained circulation of vehicular traffic via an inner ring road; in so doing, the city's post-war economic prospects would be boosted:

> Industry in and around Birmingham was very much interdependent; for instance, many small firms in the city centre manufactured components for assembly in factories on the outskirts. . . . Joseph Lucas' twelve factories employed 300 vehicles on transport within the city. The average time for a cross-city journey of 5 miles was 30 minutes and an increase in their average speed from 10 m.p.h. to 12 m.p.h. would save £20, 000 a year. On this basis the saving of such an increase in speed to the total of 28,000 goods vehicles would be of the order of £2 million a year.
>
> (unpublished notes of Minister of Transport's meeting with deputation from Birmingham City Council, 21 Nov, 1956)

The economic capacity arguments for lobbying the Ministry of Transport for government over funding for the ring road were forcibly developed by Councillor Frank Price as Chair of the Public Works Committee. The Ministry of Transport reported that 'Councillor Price said that traffic conditions in the city centre were chaotic and . . . Birmingham was the hub of a vitally important industrial area in which complementary industries were linked by road and the project was justified by the needs of industrial traffic' (NA MT 122/3 Draft Press Notice, 1956). Urban ring roads were potential money-makers and a symbol of municipal vitality (Gunn, 2011, 2018). For example, although the financial outlay for Birmingham's inner ring road was costly (estimated at £15 million), notable Council officials argued that it was necessary for supporting the city's future economic prosperity (along with bringing in £1.5 million a year in rates and rents) (Shapeley, 2012). It was understandable, perhaps, that vehicular movement assumed such a level of importance for the City.

One resident of north Birmingham recalled his dissatisfaction with negotiating the inner ring road, which altered his daily routine of walking to reach his place of work in Digbeth (south of the Bull Ring), suggesting that:

> [In the 1950s] you could cut through all of the back streets and it didn't seem so far at all and it didn't seem to take long at all [to reach the city centre]. To me, it was either under or over. It seemed to take much longer . . . afterwards . . . in the mid-60s [following the building of the Smallbrook Ringway].
>
> (Thomas, born 1942, interviewed 2007)

During the later round of data collection and while standing at the base of the Rotunda, Maggie provided a rich account of how she felt a sense of disorientation at the impact of this section of the road, especially in relation to how it altered her walking route through the historic Bull Ring area. Here she suggested that it took some considerable time for people to come to terms with the newly constructed infrastructure; this was particularly noticeable in her recollections of the area around the 1964 Bull Ring Shopping Centre:

> It took time for people to get their heads around the ring road did, it did cut things into two, you know [for pedestrians], you just felt that this part [Points to map and the Digbeth area] was separate. . . . But it was so innovative, a big dual carriageway, but people just couldn't get their head around it, it just seemed so cramped, so alien, weird, and beyond that it [seemed that the part of the city] didn't count, somehow, . . . This area [points past St Martin's Church], in particular, felt cut off [when walking] and stuff and people were puzzled as to how to treat it.
>
> (Maggie, born April 1937, go-along)

While the separation of people and traffic was designed to make shopping excursions more *temporally* convenient for city centre users in both city centres, some residents perceived that the creation of inner ring roads and the patterns of flow imagined by the 'planners' actually interrupted existing spatial travel routines in unanticipated ways. Maureen, a resident who lived in the centre of Birmingham, suggested that even the seemingly mundane act of trying to cross the inner ring road during the 1960s caused a certain alteration in her perceptions and contact with this aspect of the built form, which ensured that she had to change her route through the city core:

> I thought actually the part around Smallbrook was good with the ground floor shops. Further round [the ring road], again, . . . [when I tried to walk] it cut off Suffolk Street and all around there. You didn't see people walking across there except to go to the theatre. It [the ring road] did cut off a lot of the shops. I found my own way around it in the end; it was just too much bother to get over there.
>
> (Maureen, born 1934, interviewed 2007)

Disruption was not always limited to the pedestrian: despite the rhetoric of speed and efficiency, Jenny, for example, spoke of her recollections of driving along the newly constructed Smallbrook Ringway as being slow and congested during the mid-1960s: 'I had a job in the holidays in the tyre factory and Erdington and I used to drive across the city centre [from the south-west of the city] along Smallbrook Ringway and the traffic used to be pretty horrendous but I used to have an old car and people used to get out of the way, but I had only just learnt to drive' (Jenny, born 1943, go-along).

Other concerns were raised over the segregation of people and traffic were levelled by two other critics of the time, John Tetlow, a consultant architect and planner and the then Chairman of the Advisory Committee of Birmingham School of Planning and Anthony Goss, a former senior lecturer at Birmingham School of Architecture. They caustically commented that the 'pedestrian is treated like a second-class citizen, driven down steps and ramps into subterranean passages . . . [I]t does not yet appear to have been understood in Birmingham that there is more to redesigning its city centre than keeping traffic moving' (Goss and Tetlow, in *Birmingham Evening Mail*, 8 Apr, 1965). In response to these criticisms, the then Lord Mayor of Birmingham, Alderman Frank Price, provided a typically terse counter claim: 'these critics still seem not to have grasped the significance of the problems that faced us . . . we want to solve some of our problems now! Today it is action that is required; not words' (*Birmingham Evening Mail*, 8 Apr, 1965).

Using 'space syntax' as a method to explain patterns of pedestrian movement in cities, van Nes (2001) provides an instructive visual indication of how pedestrian routes before and after the construction of both inner ring road were spatially disrupted (Figure 4.12). In this example, the thicker lines represent busy pedestrian activity in the city centre. After the construction of the inner ring road, pedestrian routes appear to have been far more contained inside the city core.

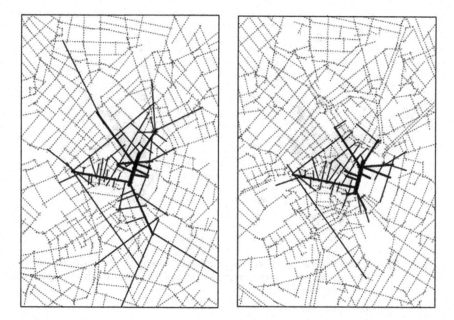

Figure 4.12 A space syntax analysis of Birmingham's city centre in 1955 (left) and 1997 (right)

Source: Reproduced with permission of Akkelies van Nes

Despite these spatial and temporal disruptions felt by some Birmingham residents, Figure 4.13 also shows how much of the city centre and how many of the shops can be reached from several sections of the inner ring road. In Birmingham, New Street, Colmore Row, and Hill Street are relatively well connected to the ring road (van Nes, 2001).

In contrast to Birmingham's inner ring road, Coventry's ring road lacks this sense of connectivity: from a functional perspective, this goes some way to explaining why Coventry's city centre, since completion of the inner ring road in the early 1970s, was described by some respondents as being 'clinical', 'dead', or, more worryingly, 'frightening' at night time (Hugh, born 1941, go-along).

Winifred, for example, suggested how 'cut off' the city centre felt because of the inner ring road:

> At first, [the Precincts] were sort of renowned all over the world for the way it was revolutionary, and they got this idea of precincts without traffic. And then they built this awful – I don't know whether they called it the flyover ring road now, whereas before, when it was first built . . . inside [the city centre] was no traffic, only buses . . . the buses and taxis, but then they built some sort of horrible flyovers, so that

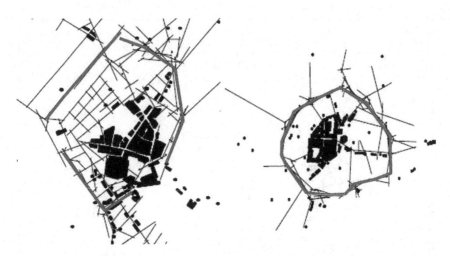

Figure 4.13 A space syntax analysis showing the connectivity of existing streets in Birmingham (left) and Coventry (right) in relation to their respective inner ring roads

Source: Reproduced with permission of Akkelies van Nes

from anywhere in the city that you look, you [had to] get [around] these awful concrete supports with lorries going over, especially round by the hospital [the north-eastern edge of the inner ring road].

> (Winifred, born 1915, interviewed 2001)

Once constructed, however, there were also pockets of support directed at the design treatment of the road. For one contemporary commentator, walking along the length of the new inner ring road evoked feelings of pleasantness:

> In walking around the . . . Inner Ring Road . . . a great deal of attention has been given to making these pleasant places for pedestrians to be in. They are not just bare areas . . . for people to race across as quickly as possible. This is particularly true of the St Chad's interchange . . . where there is a most delightful mosaic mural.
>
> (Dallaway, [representing Ove Arup & Partners],
> cited in Cowles *et al.*, 1975, page 455)

Support for the scheme was in evidence elsewhere: writing shortly after the opening of the road, Mr Cowles, of the Institute of Civil Engineers, suggested that:

> The Inner Ring Road was built with two main objectives: to enlarge the business and shopping areas of the city centre and to provide a solution

to the traffic problems in the same area. The completed road has now been in use for four years and an assessment can be made of its performance . . . with regard to the traffic problem . . . figures for all the traffic routes in the central area before and after the opening show that speeds have increased from 8 to 24 km/h at the same time that flows have increased . . . this has considerably improved the environmental quality of the area . . . the authors consider both . . . objectives have been met.

(Cowles *et al.*, 1975, page 453)

Emerging spatial practices

By 1969, when Birmingham's inner ring road was nearing completion, several letters submitted to the *Birmingham Mail* criticised the way in which pedestrian practices were being 'moulded' to the needs of motor vehicles:

When the city [centre] was being 'designed', I would have thought the traffic should have gone underneath, with many car parks, leading off the roads, then the pedestrian could have walked freely across streets from shop to shop with NO STEPS and no unnecessary miles of detours.

(A Wilkin, letter to the *Birmingham Evening Mail*, 30 Oct, 1969)

[T]he railings round the roads I deplore; they make one feel herded. The underpasses should never have been made for pedestrians. The planners should have put all traffic underneath and with people able to wander everywhere. . . . I hate going in to shop [I just] get out of the centre as fast as I can.

(G Andrews, letter to the *Birmingham Evening Mail*, 30 Oct, 1969)

Moreover, certain routines arguably emerged from pedestrian practices themselves rather than being something '*imposed*' by 'planners' upon urban walkers. Despite Manzoni's argument that pedestrians should avoid the inner ring road, or at least be encouraged to make use of the subways to traverse the road, Peter, who was working at the City Council during the late 1960s and early 1970s, recalled how he and his colleagues regularly used a perimeter section of the newly completed road:

[The ring road] did help in that [people] could walk around on the outside [of] the really busy core . . . but it was a big help if you were working like I was in Broad Street and I had got to go down to Snow Hill Station [to catch a train], I could walk down the side of the ring road. So [it] became [a] way to get around for pedestrians. [It was] very much used by commuters in that way to get into offices and that.

(Peter, born 1931, interviewed 2008)

Valerie P., for example, was hugely critical of the inhibiting influence the ring road had on her movement through town during shopping expeditions. And yet she also pointed out how she used to take advantage of the historic (and largely unaltered) back streets (such as Needless Alley, Canon Street and Cherry Street) to move from one area of town to the next.

In Coventry, the pre-and post-war reconstruction radically altered many of the existing street patterns in the city centre. However, with the very noticeable exception of the visually and psychologically intrusive inner ring road, Birmingham's street patterns largely remained relatively unchanged. For several respondents, therefore, ways of (re-)negotiating their way around the newly reconstructed city became habitual, because they will have had to do this several times as the changing reconstruction activities affected route availability. Valerie P. suggested that moving through the city centre became 'second nature in the end':

> The underpasses were the only place you could cross the road at certain places; it wasn't really a detour, though, but it did mean that you had to get to a junction where they went underneath . . . you see at Corporation Street and Bull Street there were underpasses; I used to hate those, but you had to do it to get from one side of the road to the other, you know. It became second nature in the end [moving through the city centre], otherwise you took your life in your hand!
>
> (Valerie P, born 1932, go-along 2012)

Similarly, for Steven, while the underpasses were functional – though unattractive – places, using them became part of his daily routine:

> We welcomed them – [you were] safe from the traffic. No, I think, . . . looking back to when I was at Lewis's, in the '60s . . . the underpasses were good because you didn't have to run the gauntlet [of traffic] any more but you would plan your route so once you got used to the route it became habitual. It was only when they became intimidating – [when] ruffians [were] in them, mugging people that people started to avoid the underpasses and it became chaotic then to avoid the traffic [at ground level]. People tended not to go or make sure when you were walking with someone.
>
> (Steven, born 1949, go-along)

There are certain similarities here with the narratives elicited from the Coventry respondents. There were general comments over the significant levels of destruction wrought by compulsory purchase and the clearance of slum-type dwellings in Coventry; however, most Coventry respondents conveyed a sense of 'growing-up' and 'learning to live' with the inner ring road and its associated infrastructure. Rather than something that was unsympathetically imposed on the lives of inhabitants, people found their own ways of coping with, or adjusting to, the situation. Again, Valerie G. provides a telling narrative of the difference between her father's perspective of the inner ring road and her own view of just coming to 'accept it' as being part of her routine:

> My Dad did moan about because he said that every town should have a centre point at where traffic could cross [and] I can remember . . . the bits that were built first was London Road and by the police station and St Patrick's and erm one of things I can remember when I was coming into town there was excavated out ready for it. And I remember looking and wondering about it, but . . . you just accepted the ring road and the underpasses and overpasses was the word – because there were different ones – you took in your stride. I think there is a case to accept things . . . move on no good getting rigid in any outlook really, no matter what it is. . . . You have got to keep an open mind around you in your life accepting different things.
>
> (Valerie G., born 1951, go-along)

Conclusion

As with other UK cities, both Coventry and Birmingham had to struggle with contemporary anxieties regarding the increase in motor traffic in the 1930s that had generated traffic confusion and find suitable spaces that separated traffic, pedestrians and formal civic spaces, shopping areas and parks. For both cities, plans to alleviate this situation had started before the war, but ideas of how to assuage these and other issues were further informed by the development of techniques to analyses traffic censuses; Alker Tripp's (1938, 1942) ideas on precinct design based on his experience as a senior policeman and visits to the USA. The requirements as set out in the *Design and Layout of Roads* (MoWT, 1946) and Buchanan's (1963) later recommendations of pedestrian-vehicular separation which quickly became hugely influential.[7] Both cities also made a distinction between two traffic types – through traffic and local traffic – and that, once through traffic was allowed free passage by providing ring roads, local traffic would be siphoned off via distributor roads therefore segregating vehicles from pedestrians.

More so than with Coventry respondents, however, there was a particularly strong feeling coming from Birmingham residents that the rate and the scale of rebuilding, and the 'blocky' Modernist feel of some of the individual post-war buildings, were not necessarily 'in keeping' with the city as they had remembered it as children and young adults. There was an implied reluctance to accept such change. This tended to be amplified when these memories were attached to certain familiar landmarks, personal haunts and buildings that they fondly remembered, which were being removed to make way for new developments (e.g. the construction of the Bull Ring Shopping Centre on the site of the Bull Ring market area). In contrast to the prevailing desire to improve speed of movement, it could be argued that pedestrian practices were expected (or even encouraged) to 'fit' within a 'planned for' hierarchy of flows. Both contemporary sources and interview responses suggest that the implementation of these huge and costly highway schemes caused significant temporal and spatial disruptions to established pedestrian routines (Gunn, 2011, 2018).

If anything, feelings of disruption were experienced perhaps more sharply by some Coventry residents. One reason for this might be that the separation of the inner ring road from this inner circulatory road (Corporation Street) and its realignment as an encircling ring ensured that there was sufficient space for shopping and civic zones, with other areas being loosely zoned for 'light industry' or 'clubs'. Crucially, however, the separating-out of functions (or 'zoning') also meant that the legibility did not necessarily correlate well with existing movement patterns of movement. Gibson gave considerable thought to the design qualities of the city centre, but it is questionable whether the same attention was given to those areas around the inner ring road, he perhaps believing that it was sufficient to control the redevelopment that would result from the construction of arterial roads and zoning.

In contrast, Birmingham's inner ring road managed to maintain more of a connection with the existing street pattern within the city centre, and the area within the ring road was not so sharply differentiated by land use (although there were some well-established 'clusters' of uses). Some Birmingham residents – particularly those spoken to during the go-alongs – also reported feelings of intrigue, especially towards the newness of the ring road, and occasionally people also reported using elements of the ring road for more languorous and/or leisurely pursuits rather than the need for speed and efficiency promoted by the City Council (and the Ministry of Transport). The narratives taken from the residents of both Birmingham and Coventry also speak of how pedestrian practices 'emerged' from the implementation of planners' ideas rather than notions regarding efficiency and unfettered circulation – once, of course, people became used to moving around their new environments. Chapter Five briefly outlines the approaches taken to 'dealing' with the elements of the modernist townscape following as both cities seek to regenerate the city centre.

Notes

1 Note, however, that local interest groups such as the Chamber of Commerce proposed several 'radial' plans for Coventry, though not altogether seriously.
2 Sydney Larkin's son, the mid-twentieth century poet, Philip Larkin (1922–1985), spoke of his childhood home in Coventry in his poem, *I remember, I remember*. He describes a visit to the city and caustically suggests that his early experiences in Coventry were 'unspent', saying to his travelling (rail) companion, 'as though you wished the place in Hell' (Larkin, 1954).
3 Holford replaced Abercrombie as Professor of Town Planning at University College, London, before going on to be the president of the Town Planning

Institute (1953–1954), and of the Royal Institute of British Architects (1960–1962). He is perhaps most famous for his Paternoster Square, adjacent to St Paul's Cathedral and the South Bank complex (see Cherry and Penny, 1986).

4 There is some commonality here with other examples of public feeling reported in other midland towns such as Wolverhampton, where the proposed new civic centre resulted in complaints about the perceived prioritising of bureaucrats' offices at the expense of housing (Larkham, 2002).

5 Only those bombed buildings such as the Cathedral itself, like Plymouth's Charles Church or Bristol's St Peter's Church, in part because they lay outside the main plan area, were retained as enduring evidence of the war damage suffered by the city (see also Larkham and Adams, 2016).

6 In June 1956, Coventry City Council suggested that 66.8% of the 'men' employed either in general engineering or in vehicle manufacturing and repairing compared to national figure of 15.7% (Public Relations Department, Coventry City Council, 1958).

7 For an appraisal of Buchanan and planning see Gunn (2011) and Bianconi and Tewdwr-Jones (2013).

5 New model cities

Introduction

What happened to the post-war built environment in Britain, as elsewhere in the developed world, especially between the late 1950s and early 1970s, continues to affect the daily lives of many people directly and indirectly – in fact most of those who live in or visit those towns and cities that were rebuilt. The underlying driving forces of that profound physical transformation included well-meaning, patrician values and assumptions of the newly created welfare state and a belief in the loftier virtues of male-dominated planning (whether this was actually undertaken by planners, architects, surveyors or engineers). There were other factors, of course: the beguiling influence of European modernism on groups of architectural practitioners and particularly students, a passionate repudiation of antediluvian Victorianism and its associated ills and an unfettered belief in modernity, efficiency and progress. 'Opportunity' was also clearly a factor: for property developers to profit, for the fast-growing property investment companies to manage their portfolios and for local authorities to secure environmental improvements and reposition themselves in a fast-changing post-war socio-economic context. By the late 1960s, however, especially following the Ronan Point high-rise disaster of 1968, widespread public and professional reaction to the physical form of much of the post-war rebuilding was palpably setting in. By the mid-1970s, amid another worldwide economic crisis sparked by Middle East wars, the era of reconstruction was in sharp decline. Such a crisis of confidence cast doubt on the supremacy of the architect and the validity of his (as they were predominantly male) alliance with planners and others involved in the production of the built environment.

The 1970s were, as many would attest, a low point for theoretical and practical planning, with several important studies emerging which cast doubt on the role of the planning expert. 'Real' planning, it seemed, was interpreted by some as being very distant from the 'Olympian', top-down perspective of the wartime and immediate post-war years; a damning interpretation further compounded by rapidly rising environmental concerns regarding humankind's impact on the environment (Platt, 2015). At its worst, planners become depicted as manipulators of local views, and if the underpinning of this approach is pluralism, then the role of the planner to influence changes became minimal: in this sense, as Hall (2014) points out, planners 'lost' any claims to mystical power. For some, like Wildavsky (1973), planning had become so diffuse that it was almost meaningless. Consequently, new ideas to 'bridge a gap' between planning strategies and the physical and social world were being expounded (Galloway and Mahayni, 1977, page 68). These were stinging critiques, and this is much-covered territory in planning theory and history. Yet questions remain that continue to resonate: some argued that planning theory should avoid all prescription; it should stand outside the planning process and analyse the subject more closely and reflect more critically on the historical forces. By the late 1980s, there were other planning concerns following the collapse of the Soviet Union and the Eastern European People's Democracies, re-unification, the rise of global concerns about climate change, new social movements, terrorism, insurgency, deindustrialisation and globalisation, and the list goes on. In planning terms, Harvey (1989, page 66) suggests that these new ideas – post-modernism – represented a break from 'large-scale, metropolitan-wide, rational, efficient plans', in favour of 'collage, [and] ephemerality of architectural styles'.

Modernist top-down state planning, it was argued, represented the carrier of misplaced Enlightenment principles of rationality, science, progress, control and regulation. Instead, as Sandercock suggested, more emphasis should be placed on listening and responding to the voices of those outside Western visions of modernist planning discourse; the 'insurgent histories' (1998, pages 1–2). The voices of the communities and histories of those minority groups should be listened to; this is a call for planning to acknowledge its 'dark side' (Ward *et al.*, 2011, page 248). However, there are problems with this interpretation. Sandercock manages to imply that all official planning has a secret significance and that this is inherently about regulation, control and order. Such a view tends to lend credibility to rather a paranoid version of 1990s post-modernism: planning is forever reduced to being little more than the facilitator of development, which creates 'dystopian' 'edge' cities, unfettered suburban sprawl and unregulated development, 'privatopias', fortified spaces and commodified heritage sites, all situated within a world of economic restructuring (Tewdwr-Jones, 2011).

In the UK, for example, the City Challenge scheme and the Single Regeneration Budget – key 1990s planning concepts – abandoned the fixed location (land-use approach) in favour of open competition between cities for government money. This was advantageous for some, of course: 'pro-growth' cities, especially those that had suffered the ravages of deindustrialisation, brought forward some very high-profile schemes. Hall (2014) notes that architect-planners, for example, embraced the appearance and decorative role of café-culture cities, while pointing out that the treatment of public spaces came largely at the expense of spaces where 'ordinary' city dwellers lived and worked. Moreover, the planners' well-intentioned, albeit unfulfilled, visions of efficient post-war city centres could be read as being emblematic of 'failed' modernist-inspired planning ideas, which needed radical reconfiguration to fit with post-modern times. Following the onset of economic downturn and recession during the 1980s, and again after the global economic crash of 2008, cities faced an opportunity to remove what some had already identified as troublesome elements of the post-war built form.

Repackaged landscapes

The post-modern 'turn' focussed planners' attention on the qualitative experiencing of place. The human body became more important in this context. The common complaint is that planning has been very effective at classifying people as 'agents', members of classes, or economic units that can be aggregated to areal statistics, but rarely been considered as embodied beings. And if we accept the very broad and very general proposition that cities are becoming increasingly 'plugged-in' to the global space economy, then urban spaces have a wide variety of different 'body types'. And living in a globalised urban environment means having daily encounters with people of diverse body weights, shapes, sizes, colours, ethnicities, social-cultural backgrounds, gender identities and lifestyles (Short, 2014). In recent years, a great deal of work has explored the *flaneurial* experiencing of place. For example, drawing on the ideas of Baudelaire, Benjamin, de Certeau and others, much of this work uses *walking* as an approach to understand place, particularly as many cities have become presented and used as commodified spaces of consumption replete with festivals, activities and post-modern architecture. As Degen (2017) reminds us, planning is intimately and unequivocally associated with a long-standing historical desire to 'aestheticize' places through planning and design activity. One recently rehearsed argument is that such interventions work to expunge unwanted, noxious, undesirable uses and, perhaps more controversially, remove people (rough sleepers, skateboarders, etc.) who are deemed 'out of place' and who do not conform to a particularly aestheticised vision of place. In such an interpretation, urban designers, regeneration bodies and planners manipulate the sensory environment to conform to the dominant design codes. In this process, the individual and local 'sense of place' is eroded, and the affective, emotional and sensory qualities of place are being commodified.

Post-war modernism

The architecture of the post-war reconstruction period in Britain has rarely assumed a significant level of public, professional or academic praise. However, in the UK, champions of 1950s / 1960s architecture – and post-war conservationists such as the Twentieth Century Society – have been instrumental in promoting a considerable academic and public reassessment of post-war modernist built form (see, for example, Harwood, 2015). Yet urban policy-makers in most UK cities – including Birmingham and Coventry – have been keen to rapidly eradicate or reconfigure much of the built form of the 1950s / 1960s. The architectural style of brutalism, which lasted roughly from the 1950s to the mid-1970s, has generated a burgeoning reappraisal in the recent past (see, for example, Calder, 2016). Harwood (2015), for example, offers to give us a dense and authoritative exploration of English architecture from 1945 to 1975. Her assessment goes some way towards correcting the common misgivings and collective memories about the unsustainable nature of much post-war construction, the connections between modernism and the plundering of historic towns and city centres and the top-down, bureaucratic and large-scale nature of many state-sponsored planning projects. The buildings discussed in this account unsettle the general perception that this period was solely about the architecture of monotonous boxes. In sum, this serves as a powerful reminder on the diversity of styles and approaches used by modernist architects. This is shown particularly well in the reconstruction of Exeter and Plymouth, where modernist-inspired boxy forms were clad in familiar red brick or Portland stone and, in some cases, a veneer of 'stripped' classical style (Figure 5.1).

On a more intimate level, Orazi's (2015) *Modernist Estates* takes an inside look at remarkable and sometimes controversial municipal housing estates in Britain and examines the continued impact they have on the lives of contemporary residents. It presents an overview of the buildings and architects, considers the historical and political context and explores what it's like to live on a modernist estate today. Through interviews and photography, this

Figure 5.1 Dingles department store, now House of Fraser, at junction of Armada Way and Royal Parade, Plymouth (*c.* 1949–1951) (left) and Art Deco-inspired banking hall of Lloyds Bank, Exeter High Street (*c.* 1950) (right)

Source: Authors' own collection

account offers a personalised insight into the sense of emotional attachment people have towards living in significant British post-war buildings such as the Barbican and Park Hill. In a comparable way, as Degen and Rose (2012, page 32) note, in their account of go-alongs with residents of modernist Milton Keynes, that paying attention to individual multi-sensory encounters in this way can help bring into view the 'liveliness' of urban space. This results from the interweaving of things traditionally considered 'immaterial' (recollections, feelings, emotions) with tangible material artefacts and sites (e.g. buildings, monuments and infrastructures).

For a range of reasons, definitions of the 'historic environment' have broadened very considerably in the post-war period, in both public views and quasi-legal systems, such as (in the UK) listing buildings and designating conservation areas. Within the context of planning and the historic environment – including the post-Second World War townscape – a dichotomy exists between preserving the past for its inherent value and the desire for regeneration in response to changing socio-economic and environmental values (Figure 5.2). Sandercock (1998, 2003), for example, refers to how certain modern urban planners have become 'thieves of memories' and how residents' recollections of everyday memories inscribed in urban space are 'stolen' because of modernisation and ongoing management of the built form.

There are subtleties to consider with this general line of argument. Writing in the context of post-Second World War Japan, Yoneyama (1999) questions how acts of remembrance of war can serve as complicated, multiple and contradictory narratives. Likewise, Lagae (2010), for example, shows how the meanings attached to urban sites, memorials and buildings, officially designated by the International Council on Monuments and Sites (ICOMOS), are re-appropriated as *lieux de mémoire* by residents, which, at different times, clash – but may also align – with official forms of collective heritage. Similarly, for Till (2012, page 3), in her discussion of 'wounded cities . . . steeped in oppression', individuals and social groups can experience sites differently to the sanitised, commodified and/or distorted versions of the past often enshrined in certain official redevelopment initiatives. Older individuals' repeated engagement with nostalgic recollections of their youth can be an important facilitator for positive aging. In this sense, people's perceptual memories of different sights, sounds, tastes and touch can provoke emotionally significant memories in unusual and unpredictable ways (Adams and Larkham, 2016). It is also recognised that tangible and intangible aspects of the environment also have an 'agency' that exerts an influence on humans in unpredictable ways through the capacity to arouse memories of what might have been consciously relegated to history.

The new built environment of bombed cities serves as a memorial on a large scale, but more specific memorials exist at many scales in bombed cities. These range from shrapnel scars on buildings, deliberately retained churches, to plaques, to the first bomb or to the long-range cross-channel

Figure 5.2 Exeter, retention of chapel in 1970s shopping centre
Source: Authors' own collection

shelling in Dover. Most significant, perhaps, are the shells of bombed churches deliberately retained as memorials (Figure 5.3). However, as the people who experienced the bombing age and die, these ruins and spaces are appropriated by new users and uses, and their roles as memorials change.

Instead of the 'irreversible past', much recent research focuses on a 'persisting or haunting past', and despite the diverse physiological and biological changes that happen continually to the human body, memory has a durability and the capability of returning. This is a point acknowledged by DeSilvey and Edensor (2013, page 472), in their discussion of lost geographies associated with 'mundane' modern urban 'ruins' and how the materiality of urban spaces can help to stimulate an individual's 'imaginative engagement with the past', even when memories are seemingly suppressed (see also Drozdzewski *et al.*, 2016). The persistence of memories varies according to age and personal disposition, of course, and individuals may also choose to forget as a way of dealing with feelings of physical and mental dislocation. They may seek to avoid distressing episodes of the

Figure 5.3 St Thomas, Birmingham, now a 'peace garden'; tower of St Alban, London, converted to office and accommodation; Holy Rood Church, Southampton; Charles Church, Plymouth, marooned on traffic roundabout

Source: Authors' own collection

past and discard physical *aides-mémoires* that might provoke unfavourable memories. In this sense, perhaps, memory has a functional quality and humans only record and remember things that are important for our survival: to have a perfect memory would hinder the day-to-day existence; people may avoid sites/artefacts to subdue injurious unsolicited recollections (Muzaini, 2015).

Encouraging studies continue to emerge across a range of disciplines and contexts, which highlight the multifarious embodied engagements people, have with different urban sites and how the material surroundings, buildings, artefacts, memorials and even absent sites can provoke unexpected narratives of place. Reflecting on and extending these instructive lines of inquiry, therefore, attention now turns to the various ways in which Coventry and Birmingham residents recall the post-war attempts the more recent moves to regenerate the two city centres. Before this, however, it is necessary to sketch out some of the recent regeneration attempts experienced in both cities.

Birmingham's global ambition

Parts of Birmingham's post-war built environment continue to assume a certain importance in the cultural identity and memory of the city. As Moran (2010) pointed out, Birmingham's once gleaming and modern city provided an unusually (and peculiarly) glamorous backdrop for the 1973 film *Take Me High*, in which Cliff Richard plays a merchant banker mistakenly relocated to the industrial midlands rather than the more urbane surroundings of Paris. In one early scene from the film, he glides serenely beneath the then newly constructed 'Spaghetti Junction' (Gravelly Hill Interchange – built in 1972) along the city's historic canal system in a hovercraft to arrive promptly at his destination. Despite his encouraging experience of navigating his way through elements of the city's post-war built form, the film's star went on to criticise Birmingham's tortuous, confusing and congested road system (Figure 5.4). Such a perspective emerges elsewhere, with Moran (2010) pointing out that despite febrile public enthusiasm for the opening of 'Spaghetti Junction', there were also portentous warnings reported in local newspapers during the early 1970s of the over-bearing and confusing nature of the highway infrastructure.

It is arguable that the post-war physical landscape fell some considerable way short of public aspirations, and, according to one outraged critic, '[o]ver-enthusiastic developers tore out much of the city's heart and replaced it with lumps of featureless concrete' (cited in Birmingham City Council, 1989a, page 143). Others, such as Upton (1993, pages 184–185), suggested that 'the shopping focus and traditions of Birmingham were changing [. . .] no one knew where the centre of Birmingham was', while the Bull Ring Shopping Centre had 'subways and broken escalators', which made the public less inclined to take the plunge down into the Bull Ring. Planners had to rethink the merits of single-use zoning, while within the city core, many

Figure 5.4 'Stranger in Paradise'; a public representation that conveys a sense of a Birmingham as a city dominated by highway infrastructure

Source: Authors' own collection

of the pedestrian subways were filled in from *c* 1990. Ground-level crossings replaced the subways, compromising the free flow of traffic, but facilitating pedestrian flow to a certain extent.

During the 1980s and early 1990s, under the leadership of Richard Knowles, Birmingham City Council sought to pursue a new future for the city; one based on grand development projects and aggressive place-promotion initiatives. These were an attempt to reverse the city's image, which was centred on the primacy of the motorcar (for a more in-depth review, see Hubbard, 1996; Loftman and Nevin, 1996). Most of this reforming enthusiasm focussed on the legacy of the radical physical transformation that occurred between 1947 and 1973, characterised by the modernist redevelopment and overwhelming highway schemes. Henry and Passmore (2000) point out that a series of policies, which implemented urban regeneration within Birmingham city centre, represent an attempt to reverse images of parochialism, militancy and poverty.

In 1988, Birmingham City Council decided to make drastic changes to the layout of the inner ring road. The City Centre Symposium, or High-bury Initiative, held in March 1988 suggested that the inner ring road should be retained but that its role as a by-pass around the centre be transferred to the middle ring road; the inner ring road would then be 'downgraded' to become another 'city street' (Birmingham City Council, 1989a, 1989b; Corbett, 2004). Support for the City Council's pro-growth strategies and city centre prestige projects found expression through the hosting of city centre symposia (called the Highbury Initiative Symposium 1 and 2 respectively). The Symposia underlined the importance of maintaining the economic vitality of the central area, and this meant making it highly accessible by all forms of traffic. However, the unanimous conclusion of the Symposia was to give more thought to the role of pedestrians in the city centre. Furthermore, the Highbury Initiative represented a key moment in the city's rejection of its modernist achievement (Higgott, 2000). In part, this was in response to the inner ring road's subjugation of the pedestrian in favour of high-speed vehicular traffic. But perhaps more importantly, it was also because the ring road had quickly come to be seen as a 'concrete collar' – 'a physical and psychological barrier inhibiting growth' of the city core business and retail district (Birmingham City Council, 1989a, page 18). One expression of this belief was the decisive rejection of aspects of post-war architecture and urban form: *The Birmingham Urban Design Study* (Tibbalds *et al.*, 1990), for example, also suggested ways in which the city could overturn its image of an unfathomable concrete jungle by making it legible, navigable and by prioritising the pedestrian over the motorcar.

A narrative of regeneration is familiar to Birmingham. In the 1950s and 1960s, the wider West Midlands region underwent change on an unprecedented scale, as Birmingham reaped the benefits of manufacturing-related economic success. However, in retrospect, the specialised nature of the region's economic base made it increasingly vulnerable to structural economic changes. Hence, the protracted recession of the late 1970s and 1980s had a dramatic effect on the region, with employment in the dominant manufacturing sector declining rapidly, with some 191,000 jobs lost in Birmingham between 1971 and 1986 or 29% of all employment, and manufacturing employment was virtually cut in half with the loss of 150,000 jobs (cited in

Flynn and Taylor, 1986). The physical form of the reconstruction areas was also altered during this period: in the five 'new towns', for example, many of the city's large collection of tower blocks were demolished (P. Jones, 2004). This point was recently re-affirmed by local councillor, Robert Alden who declared that 'we will demolish every council high rise tower block in Birmingham' (Elkes, 2018).

In a sense, the pro-growth strategies developed under the then council leader Albert Bore (1999–2004 and 2012–2015) were tempered by the recognition that the needs of Birmingham's disadvantaged groups and most deprived areas were not being met by the physical rejuvenation of the city centre area. Moreover, the designation of these central areas demonstrated the City Council's determination to revitalise the city image. This view represents a shift away from a sense of heroic modernism that had prevailed in the post-war redevelopment of the city to a more pragmatic, prosaic and piecemeal approach. Rather than utopian or visionary frameworks and redevelopment according to idealised blueprints, new development at the turn of the century tended to focus on neatly packaged, individual projects and areas.

Birmingham has been exceptional in the manner that it has explicitly acknowledged the intimate link between design and investment in its long-term vision as set out in the Highbury Initiative, claiming that 'the attraction of major organisations together with local growth, could provide the spur for exciting architecture of high quality, which the city currently lacks' (Birmingham City Council, 1989b). These progressive statements, which superseded all previous plans, made a clear case for the role of design in the pursuit of economic and urban regeneration and re-emphasised the City Council's continual commitment to environmental improvement (Punter, 1994).

Individual projects included the well-known developments like the National Indoor Arena (NIA), the International Convention Centre (ICC), Birmingham Symphony Hall and a series of retail and leisure complexes such as Brindley Place, the Mailbox and the revamped Martineau Galleries. In addition to these projects, the City Council designated a Chinese Quarter (Chan, 2005), cleaned up the canal system and pedestrianised large sections of the city centre (see also Pollard, 2004). Alongside the leisure, retail, educational and cultural developments, another facet of the City Council's regeneration effort has been to attract more people and especially young, skilled people, to live in the city centre. The City Council's 'City Living Initiative' has promoted the construction of accommodation geared towards high earning, often single, mobile professionals and investors (Pollard, 2004). Millennium Point (home to science, technology and educational projects) was completed, the largest Millennium Project outside of London and was intended as a catalyst for the regeneration of the Eastside quarter (Figure 5.5); and the redeveloped Bullring re-opened in 2003.

Figure 5.5 Birmingham's Millennium Point and Eastside City Park (2012)
Source: Authors' own collection

Emphasising the attractiveness, necessity and opportunity of redevelopment, the images and documents promulgating Birmingham's more recent regeneration initiatives deliberately juxtaposed the deficiencies of the postwar city (typified by over-bearing highway engineering projects) with the promise of a more socially inclusive, profitable and sustainable townscape (Hall and Hubbard, 2014). Moreover, as Miles (2002) suggests, this process of regeneration is not simply an act of physical renewal, it also relates to building a new visual image, or symbolic economy. Hence, the visual narratives of regeneration underlying certain city centre flagship projects are often accompanied by a new politics of erasure (Higgott, 2000). Here, the new cannot be built until aspects of the old are first symbolically dismissed before the area is redeveloped, re-assembled and re-ordered (Campkin, 2014). This was certainly true for the Central Library (Larkham and Adams, 2016).

Despite this 'boosterist' rhetoric, one of the crucial political and economic questions regarding this extensive regeneration concerns which constituencies are included and excluded from participating in these new residential and consumption spaces in Birmingham. Zukin (1998) also acknowledges the divisive inequalities this creates, recognising the existence of populations

who are of little interest to the corporate planners and property developers who cater to mobile and young, mainstream consumers (Birmingham
City Council, 2011). McEwan *et al.* (2005) develop a similar account based
on the importance of Birmingham's ethnic communities, as once marginalised groups, being a source of economic and social vitality and innovation
for the city. The city's *Big City Plan: City Centre Masterplan* (Birmingham
City Council, 2011) highlights the importance of creating a city centre that
is 'connected'; the type of urban centre that is responsible for coordinat-
ing globalisation by acting as a strategic hub in the global space economy,
emphasising the hypermobility of people, goods, capital and information:

> Central to Birmingham's success in the future will be the strength and
> sustainability of its economy which will need to be robust and diverse
> enough to perform alongside European and international competitors.
> Creating a strong post-industrial economy driven by a range of sectors
> including knowledge, technology, research, science and local services
> will be key.
>
> (Birmingham City Council, 2011, page 8)

The plan also envisages city-centre uses expanding across the broken
concrete collar into Digbeth, Eastside and other zones. Most recently,
the announcement of a high-speed rail station at Curzon Street (HS2) has
given impetus to this rebuilding of the east side of the city centre. In their
analysis of how Birmingham's entrepreneurial developments sought to
become plugged into the global 'space of flows' (Castells, 2000), Henry
and Passmore (2000) speak of the apparent failure of symbolic sites like
the ICC to 'tap-in' to local people economically or culturally (see also P.
Jones, 2008). While certain post-war structures have been retained and
protected,[1] there is certain creative value associated with the destruction,
or selective erasure, of elements of the post-war townscape (Colby, 1964);
or what Schumpeter famously termed the idea of 'creative destruction'
which 'incessantly destroys the old'. Some of Birmingham's notable post-
war buildings, though praised by several respondents – at least when first
constructed – have been threatened with demolition or have already been
demolished. Prominent 'corpses' (to borrow Le Corbusier's term) within
the line of the inner ring road include the original Bull Ring Shopping
Centre, replaced by a new mall, which has re-established the pedestrian
connection between High Street and Digbeth. Madin's AEU building
situated on the corner of Smallbrook Ringway has been replaced by
the city's tallest hotel building, his RIBA-award winning Post and Mail
building – felt by some architectural critics to be his best – was demolished in 2002 (Foster, 2005). His Central Library and NatWest Bank
have been demolished, despite the efforts of Elain Harwood and other
champions of the Twentieth Century Society to protect them (Clawley,
2011; Larkham and Adams, 2016). Roberts's Rotunda has fared better:

it was threatened with demolition, but by *c.* 1986 was reprieved and Listed. Yet it has also been gutted, reclad and is part of the new Bullring shopping centre hangs awkwardly from it. Perhaps most strikingly, the Masshouse Circus junction of the inner ring road has been levelled, thus allowing the outward spread of development to proceed in to the east side of the city centre.

The design of these new 're-packaged' landscapes is seen as a crucial instrument in the orchestration of a new 'entrepreneurial' urban image. For some, the perceived demands of the global economy take precedent over the experiences of local residents as cities become increasingly sanitised, commodified and distorted. Town and city centres are under increasing pressure to 'perform' as marketable commodities, with effective planning required to promote competitive town centre environments on an international, national, regional or sub-regional basis (DCLG, 2012). Birmingham's *Big City Plan* (Birmingham City Council, 2011, page 25) continues and develops this narrative of selective dismantlement of the city's post-war legacy, arguing that any further expansion of the city core north and south should look to 'break through' the former inner ring road, thus generating a flourishing walkable 'open space network' for city centre users (see also, Birmingham City Council, 2011) (Figure 5.6). For

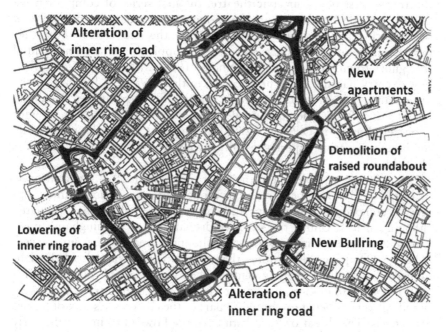

Figure 5.6 Recent changes to Birmingham's city centre
Source: Crown copyright reserved

Birmingham to 'operate in a more competitive and globalised world', the city council notes that one of the key factors for its future success 'will be the strength and sustainability of its economy' (Birmingham City Council, 2011, page 8). The substantial (a reportedly £600m) refurbishment of the post-war New Street Station, together with the opening of Arena Central on the site of Madin's library continues the narrative of city renewal as it seeks to 'bring major economic benefits, helping economic growth by creating new jobs and stimulating regeneration' (*Birmingham Mail*, 15 Dec, 2012). In 2018, this approach was bearing fruit: Birmingham was voted a 'Destination of the Future' by fDi Intelligence (*The Financial Times*), as the city became one of the highest performing cities in all of Europe and the third most globally influential city in the UK (after London and Manchester) (Mullen, 2018).[2]

Changes to Coventry's built form

As with Birmingham, the sharp contraction in manufacturing employment base during the 1970s and 1980s ensured that the fortunes of Coventry's city centre took a downward turn (see Campbell, 2004, 2006). In an early account that presaged later criticisms, Taylor (1969, page 4) suggested that 'Coventry's problem' was its pioneering qualities: 'the first comprehensive town plan, the first large-scale pedestrian shopping centre, the first post-war civic theatre, the first series of comprehensive schools outside London, [but] the pioneer of one decade runs the risk of appearing the has-been of the next'. By the mid-1990s, the rather unwelcoming and bleak character of the shopping precincts represented the 'ghost town' (Healey and Dunham, 1994; Millet, 2000) described by Coventry-based band, The Specials, in their 1981 UK hit song of the same name. The precinct was beset by significant problems stretching back to the 1960s. The Specials also had a number 7 UK chart hit with 'Concrete Jungle' in October 1979; this song was reportedly inspired by the largely negative experiences of growing up on Coventry's post-war Hillfields estate.

Following the completion of the inner ring road in 1974, major aspects of Gibson's plan had been realised; though, of course, not without significant compromises, iterations and changes, especially to the designated shopping spaces. Unlike the selective dismantling of Birmingham's post-war built form, elements of Coventry's shopping area have been preserved. While (2006, page 2409) notes that redevelopment has also helped to improve links within the city centre: when it (re)opened (in 2003), at a cost of some £38 million, the Lower Precinct sought to attract high-street brands. It is now being promoted as part of the city centre's mix of leisure attractions. The threat of 'spot listing' of the Lower Precinct in the early 1990s was enough to ensure that there have been no wholesale alterations made to the shopping area. It should also be noted, though, that the Upper and Lower Precincts do not fall within the City Council's designated conservation areas (Figure 5.7).

The Lower Precinct was regenerated between May 2002 and 2003; the Rotunda cafe was restored and a canopy over the Lower Precinct was inserted (see Figure 5.8).[3] Furthermore, the Rotunda café has been granted Grade II Listed status by English Heritage / Historic England. As While (2006, page 2407) notes, 'Coventry's Lower Precinct shopping centre by the mid-1990s was regarded by post-war conservation interests as one of the best, if not the best, surviving example of 1950s / 1960s town-centre retail architecture'. From an architectural perspective, Demidowicz (2002) argues that the Precinct is renowned for its multilevel planning and 'Festival of Britain' modernist design. There is also some resilience embedded in the plans and designs conceived by Gibson and his team. Many of the elements of the 1950s / 1960s shopping centre have survived for about seven decades, though changes have been made to the overall coherence of the design.

For example, additions were made to the Upper Precinct (e.g. the creation of the escalator leading into the West Orchard[s] Shopping Centre).

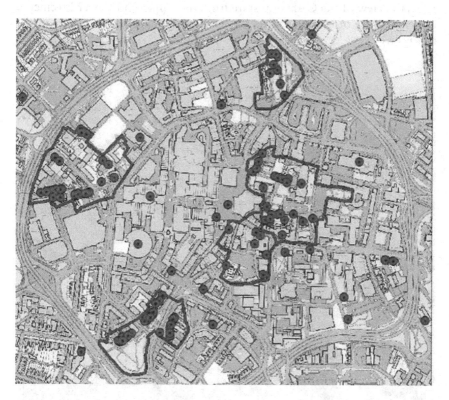

Figure 5.7 Statutorily listed buildings and conservation areas in the centre of Coventry
Source: Crown copyright reserved

And perhaps one of the most significant alterations to the Upper Precinct occurred in 1978 when pairs of symmetrical curving stairs giving access to the upper levels were removed. At the Broadgate (east) side of the Precinct, the stairs were replaced by a ramp in a perhaps belated effort to improve pedestrian and wheelchair access. In some ways, this introduction has ensured that the space between the former Hotel Leofric and Broadgate House has been considerably narrowed. The access to the circular Rotunda café has been altered, but the most significant aspect of this reconfiguration is the addition of the glazed roof over the whole of the Lower Precinct which encloses the entire open area with a pitched roof on a structure that jars with the original design intention of the precinct (Figure 5.8).

Deviating dramatically from Gibson's 1941 plan for the Broadgate area – designated originally as open space and part of the axis from the Cathedral spire through the Precinct – the decision was made in 1985 to proceed with designs for a new shopping centre. The Cathedral Lanes Shopping Centre (1989, by Chapman Taylor and Partners) represents the first fully enclosed shopping centre in Coventry and encroaches on the formerly open space in Broadgate (Gould and Gould, 2016). More fundamentally, though, it interrupts the view of the Cathedral spire from the Upper and Lower Precinct, a critical feature of Gibson's ideas for the overall image and structure of the city centre.

Figure 5.8 Glazed roof in the Lower Precinct
Source: Authors' own collection

Changes have continued in the centre of Coventry. In 2000, for example, Coventry's Millennium scheme saw the City Council's Phoenix Initiative. This made explicit connections with the earlier idea encapsulated in the city's levelling stone and in Basil Spence's (1962) text, that Coventry would, like the phoenix, rise (once more) from the ashes of its destruction. The project was intended to provide a relationship between the Cathedral and the (newly extended) Museum of British Road Transport through a series of interrelated squares, revealing Coventry's ancient historic fabric and culminating in a contemporary place, celebrating Coventry's engineering and technological successes. The plan, driven forward by MacCormac Jamieson Pritchard, signified a shift away from the Gibson 'land use zoning' approach by introducing a mix of shops, offices and residential buildings into the city centre.

Elsewhere in the city, other recent developments have deviated from Gibson's earlier ideas. For example, the demolition in 1961 of the Opera House (built in 1889), used as a cinema since the war and of the Hippodrome (built in 1937, designed by W. S. Hattrell and Partners) in 2002 left the Belgrade as the only theatre in the city centre. The Hippodrome (also known as the Coventry Theatre) survived until demolition (in 2000) to create Millennium Place and the Coventry Transport Museum. The Gaumont Palace cinema (1931) in Jordan Well has since become the University's Ellen Terry Theatre. Furthermore, a new leisure and entertainment complex now occupies a large site west of Queen Victoria Road, which was zoned for light industrial use in the 1945 Plan. This includes a SkyDome Arena (built in *c*. 2000) including a large multi-purpose area, a multiplex cinema, an ice rink and associated car park adjacent to the inner ring road (Figure 5.9).

This development has been complemented by the addition of the seven-storey Ikea store (built in *c*. 2007 and designed by Capita Ruddle Wilkinson)

Figure 5.9 Ikea, Coventry
Source: Authors' own collection

(Figure 5.9) on the other half of the SkyDome Arena site next to Queen Victoria Road. Outwardly, the building's appearance and use of the company's distinctive blue and yellow panels is distinctly incongruous, contrasting sharply with the more modest timber frame buildings on the neighbouring Spon Street (and the Spon Street Conservation Area), while the scale of the building rather overshadows the surrounding street scene. Broadgate has also seen radical recent change: the 'city centre's main events space', where new paving materials, trees and seating, alongside improved architectural lighting help to bring about a sense of vitality to the city centre.

In 2008, Coventry City Council unveiled the Jerde Partnership's plan for the comprehensive redevelopment of the whole shopping centre and the area bounded by Corporation Street, The Burges, Cross Cheaping, Broadgate, Hertford Street, Warwick Row, Greyfriars Road and Queen Victoria Road (Gould and Gould, 2016). The original design intention was to retain elements of the original precincts, most notably the buildings of Broadgate and the Upper Precinct, together with the three terminal towers and the West Orchard Shopping Centre. Although the proposal sought to provide a much-needed 'shot-in-the-arm' to 'the central area's dying nightlife, by altering the post-war built form of the shopping area to fit with the mood of the times' (*Coventry Evening Telegraph*, 28 Feb, 2012), the scheme received much adverse local comment and faded away with the impact of the economic recession.[4] In February 2012, the Council unveiled its plans for the City Centre South redevelopment. The new 'blueprint' from UK Architects Benoy (architects of the redeveloped Birmingham Bull Ring), was partly inspired by the Gibson plan but also retains some of the visions included in the Jerde vision (Coventry City Council, 2012). At the time of writing, the City Centre South development proposes to revive Bull Yard, Shelton Square, City Arcade and Hertford Street by 2022 (Figure 5.10) (Lomholt, 2016).

Figure 5.10 Bull Yard, Coventry
Source: Authors' own collection

As with the Jerde plan, these more recent changes are perhaps best described as combining post-modern elements with a revival of modernist principles that took hold in the years before the global financial crisis of the late-2000s (see Fischer and Larkham, 2018). A further group of post-war buildings was listed in 2018. Duncan Wilson, Chief Executive of Historic England, said:

> The reinvention of Coventry after the Second World War and the vital role that its post-war architecture played in restoring pride and confidence was renowned internationally. These listings recognise a vitally important period in our national life, and places that have now come of age and will continue to play an important part in the evolving life of this great city.
>
> (see https://historicengland.org.uk/whats-new/
> statements/Coventry-Listings/)

The former Hotel Leofric, Woolworths building, Locarno Dancehall (now the Central Library), British Home Stores, the Levelling Stone, the Broadgate Standard, Marks and Spencer, the North & South Link Blocks and Piazza are Grade II listed. However, some have protested this decision, concerned that protection will prejudice proposed redevelopment. Councillor Jim O'Boyle, Cabinet Member for Jobs and Regeneration, said 'Historic England has proved to be very difficult to work with as development partners. They have made inward investment into the city difficult. They have gone out of their way to make sure that the development of the Upper Precinct does not happen. They are unaccountable and not fit for purpose.... This has not made redevelopment prospects easy, but we will persevere' (see www.coventrytelegraph.net/news/major-city-centre-redevelopments-risk-14462865).

Looking to the future . . .

Just as some contemporary architects saw the aerial bombardment of the Second World War as an opportunity, so we can see the alteration and demolition of the reconstruction-era buildings today. However, it would be grossly unjust to suggest that any sense of public grievance/outcry over the loss of post-war buildings should be considered uncritically without giving due consideration to the fact that times were against the retention of seemingly 'out-dated' buildings. This reason adds to the prevailing mood for change that was apparent amongst members of the public for both cities. There were also wider socio-economic forces driving urban transformation and the push for regeneration; and the sense of unpreparedness of planners/architects towards dealing with post-war 'heritage'. All these reasons have contributed to what some might interpret as the regrettable destruction of some post-war structures. Chapter Six goes on to discuss these ideas further.

Notes

1 For example, New Street Signal Box on Navigation Street, designed by Bicknell & Hamilton and W. R. Healey and built in 1964, is Grade II listed.
2 See: www.fdiintelligence.com/Locations/Europe/fDi-s-European-Cities-and-Regions-of-the-Future-2018-19-FDI-Strategy-Cities.
3 In addition, original neon light sculptures were restored, and Gordon Cullen's murals were retained and relocated in the precinct, while there have been improvements in terms of access to the upper tier.
4 See www.bbc.co.uk/coventry/content/image_galleries/jerde_plans_for_cov_sep08_gallery.shtml.

6 Recollections of urban renaissance

Introduction

Following public and press criticism about modernising urban planners and their unsympathetic removal/alteration and destruction of elements of the historic urban fabric, the collective endeavours of agents of the state, developers and preservationists have formed collaborative 'placemaking' partnerships. It is certainly the case that conservation-led development has become prominent in both city centres. However, the economic recession and ensuing deindustrialisation of the 1970s-1980s meant that much of Birmingham's post-war physical landscape came to be derided in several prominent public and professional accounts sometimes surprisingly soon after construction (see HRH Prince of Wales, 1989). Therefore, as the previous chapter described, some of Birmingham's prominent post-war developments have been demolished and replaced to make way for an urban renaissance, where visually striking post-industrial spaces dedicated to office, commercial and retail uses are designed to attract investors, tourists and consumers.

Coventry is also having to respond to pressures for the redevelopment of its 'out-dated' centre (Gould and Gould, 2016); however, much of the post-war city physical environment has been 'creatively preserved' (While, 2006, page 2416) through national protection and remains largely as Gibson (and his colleagues) intended, despite piecemeal alterations. Moreover, the city's wartime destruction continues to be mobilised by several national and civic bodies as a commemorative resource of international post-war reconciliation (Goebel, 2011). This is represented in several ways: the city's old and new Cathedrals and the University of Coventry's Centre for the Study of Forgiveness and Reconciliation, while the city is also joined by other bombed European cities, including Dresden, Kiel, Caen, Lidice, Belgrade, Arnhem and Warsaw in a commemorative network (Gutschow, 2013). This chapter draws on the local narratives to explore current local feelings about the post-war buildings, the design and use of materials, before moving on to critically discuss local people's feelings towards the latest waves of regeneration in Coventry and Birmingham and the role of personalised memory in shaping experiences of these cities.

Experiences of the repackaged city

There certainly was a prevailing sense among many respondents that the quality of materials and the style of (though largely unspecified) 'square' and 'blocky' reconstruction-era buildings were not necessarily 'in keeping' with the needs of the contemporary city. There is an obvious temptation to suggest that these narratives should be taken as being a counter or antagonistic to the attempts of the modernising planners in the 'reconstruction era' to radically reshape the city centres. Elements of the post-war rebuilding – including the Rotunda, the Central Library (Birmingham) and the shopping precincts and inner ring road (Coventry) – have been arguably woven into the collective post-war memory of the two cities, as bold declarations of modernity (see Parker and Long, 2000, 2004; Gould and Gould, 2016). Giving voice to the individual recollections of actors directly involved with the rebuilding of these two cities helps to contextualise post-war reconstruction. It helps to bring into view how the resulting new buildings and infrastructure represents 'not simply the accomplishment' of a single architect or planner (Jacobs and Merriman, 2011, page 211) who uncaringly imposes their ideas onto the level of lived space. While Brian Redknap, James Roberts and John Madin all spoke proudly of their contributions to creating 'modern' Coventry and Birmingham, they also tended to downplay recollections of their professional role in facilitating the more sweeping aspects of the rebuilding. They preferred instead to apportion blame to Manzoni, Gibson and the modernising ambitions of the city authorities and their professional planning knowledge and 'official histories'. They questioned the architectural and planning shortcomings of Manzoni and Gibson:

> [Coventry] had a congested old centre, there [was] a whole belt of poor quality factories and working-class buildings and this [belt] of outworn housing that needed to be redeveloped. . . . But some heritage was destroyed before the war, in the 1930s when Trinity Street and Corporation Street were built to help deal with the traffic. So, there was a huge amount of destruction then Butcher Row, by the Cathedral – huge numbers of timber-framed buildings. Though one of the sad things was Gibson's set-in motion the wider destruction of the city's archaeological history . . . and . . . in a way, I was also involved in it [the destruction]. We had total support; I wasn't aware of any criticism at all from the public.
>
> (Redknap, interview)

I just think [Manzoni] hadn't got the architectural concept experience to realise what you could do with a three-dimensional master plan for the city centre and so I, I've been frustrated for the last fifty years over this. . . . I just think [Manzoni] hadn't got the architectural concept experience to realise what you could do with a three-dimensional master plan for the centre of the city, I just don't think he realised how important it was to do this.

(Madin, interview)

[Birmingham] was very active and busy before the war, bustling with people and cars, and it did need some help [with reconstruction] . . . [but] Manzoni . . . wasn't interested in people or pedestrians . . . he did considerable damage to the heart of Birmingham; it needed rebuilding, but there should have been a lot of tender loving care after the war.

(Roberts, interview)

The views of Madin, Redknap and Roberts about heritage, protection and redevelopment at the time of reconstruction would have been very different to those of today. Heritage and conservation were only minor concerns for town planning – particularly for industrial cities – until into the 1960s. Nonetheless, their narratives do point to how the plans and the development of specific sites emerged out of a response to broader socio-economic forces, including, among others, the increased levels of vehicular traffic and changing housing, retail and leisure expectations. As other accounts make clear, development was also the product of personal, political or professional influences at the national and local level (Richardson, 1972; Sutcliffe and Smith, 1974). In Birmingham's case, for example, the City Council already owned many freeholds in the centre and along the proposed ring road. Prospective developers were encouraged to purchase sites, pass them to the Council and receive a lease over the enlarged area. This gave the authority considerable control and flexibility over how to shape development. However, James Roberts also points out how practical difficulties and the cost-saving ambitions of property developers, working within the parameters of the current planning and legal frameworks, influenced the design of certain buildings and infrastructure:

In the late '50s, I wanted a horizontal emphasis [for Smallbrook Ringway]. I wanted it to align with the inner ring road, almost like a city wall. It is made of pre-cast concrete, but in fact, I had a lot of thought and feeling for the John Lewis store[1] in London and that a glass and metal cladding; that would have been appropriate for the scheme. But the trouble is, I had to put up with a demoted quality [because] the developers decided that since they were involved in the thing, the materials and design became more of a cost-cutting exercise.

(Roberts, interview)

Despite the considerable changes made to the built form following the post-war reconstruction, Birmingham's industrial heritage (Birmingham City Council, 2011), is arguably championed over the city's post-war materiality.

The Bull Ring Shopping Centre has been demolished (in 2000) and replaced by a new 'Bullring' mall (opened in 2003), and Madin's prolific output is particularly threatened. Madin was not opposed to the process of urban redevelopment; however, his narrative does blend feelings of personal loss with a sense of longing for a 'better' future in a more sensitively handled and less sweeping, approach to urban regeneration:

> [My] library should remain a civic centre in the civic heart of the city [but the City Council] wants to sell the old library site for . . . new commercial buildings once they've knocked [it] down!
>
> (Madin, interview)

Most residents, however, spoke optimistically about the broad principles of regeneration efforts to redevelop Birmingham's post-war landscape following the deterioration of the physical environment in the 1970s–1980s. The new Bullring shopping centre, for example, is the city's most significant example of a building constructed (on a similar footprint) to replace the previous 1960s modernist shopping centre (demolished in 2000). Moreover, a 'flagship' scheme such as the latest Bullring, identified as being 'one of the UK's top ten retail destinations [in 2013]', attracting some '40 million shoppers per annum' (see Hammerson plc., 2013) and acknowledged as having a 'considerable beneficial impact on the profile of the city centre' (DTZ, 2013, page 13), has also been a highly visual symbol of Birmingham's wider regeneration ambitions. As Parker and Long (2004) note, hi-tech promotional imagery has been deliberately employed to emphasise the striking aesthetic qualities of the new development (Figure 6.1).

The redevelopment of the Bullring as part of a joint venture with Hammerson, Henderson and Land Securities was completed in 2003. The intention behind this project was to revitalise this area of inner city retailing after a decade in which developers focussed investment on out-of-town sites such as Merry Hill in neighbouring Dudley (Spring, 2003). The advantages of the new Bullring as outlined by the Bullring Company – established under the auspices of the developers – was to provide 'an attractive series of malls, covered streets . . . young fashions' and a 'cluster of brands' (see http://visit-birmingham.com/). This falls nicely in line with the City Council's expressed ambition, contained with its *Big City Plan*, of 'improving the retail offer for the young' (Birmingham City Council, 2011, page 9):[2]

> Designed by concept architects Benoy, Bullring has been conceived as a series of malls, open spaces, covered 'streets' and public piazzas, with the two department stores, Debenhams and Selfridges, providing an anchor on each side of the scheme. The design of Bullring's mall pattern has created a unique opportunity to cluster brands in a series of prime locations across three trading levels . . . each trading level has its own distinct personality in terms of retail mix: high street fashion and

Figure 6.1 The latest Bullring
Source: Authors' own collection

al fresco dining at Lower Level; younger fashion and lifestyle retailing at Middle Level; and aspirational fashion on the Upper Level,
East Mall.

(https:\\www.visitbirmingham.com)

In 2011, a new restaurant 'hub' was created facing the space between
the Bullring and St Martin's Church – known as St Martin's Square;
this includes Browns Bar & Brasserie, Nando's, Wagamama and Jamie's
Italian. There have been some positive responses towards the design of
the new Bullring. The removal of a section of the 'elevated inner ring,
the underpasses and access stairs', Spring (2003, page 1) argued, has
ensured that there are more ground-level entrances to shops encouraging permeability. Emery (2006, page 121) gushingly suggested that 'the
experience [of the Bullring] has shown that future regeneration projects
need to articulate an appetite for change if success is to be achieved and
sustained'.

Respondents, however, had rather mixed opinions of this latest development, especially given its emphasis on 'high-end' retail. They tended
to approve of the 'iconic' Rotunda and its recent improvement; although
fears of 'placelessness' (Relph, 1976), the 'young-oriented' and 'expensive-
looking' shops (Sylvia, born, 1937, interviewed 2007) appeared to be a

point of concern to some (see also Holyoak, 2004). This was especially the case for those who remembered the hustle-and-bustle of the Bull Ring markets of the 1950s and, to some extent, the ideas enshrined in the 1964 Bull Ring Shopping Centre:

> I've been once [since the building of the Bullring] and I thought oh, it was terrible really. . . . I didn't like it and that's why I wouldn't go back again! . . . Me and my cousin did go inside – we didn't like it. Well, it's too expensive anyway isn't it? You know, [I] couldn't afford their prices . . . they've spent a lot of money on Birmingham but at the end of the day I think they've wasted a lot of money. . . . It isn't for the benefit of the people in my opinion. [It is for] business, people don't have Birmingham like they used to.
>
> (George, born 1933, interviewed 2007)

During the go-alongs, though, other individuals spoke positively about the Bull Ring area; and despite the physical changes, it continues to represent 'their' 'playground' (Iris, born 1934, go-along). While it offers a visually appealing attractive series of malls (Birmingham City Council, 2011), it also contrasts with the Bull Ring market area of the 1950s. This was a space where they would often meet with family and friends. In the minds of other people, however, talking about the architecture and materiality of new buildings stimulated some unsettling 'ghosts of place' (Muzaini, 2015). Though James Roberts was largely supportive of having his Rotunda building nationally protected and intertwined with the recent regeneration of the Bull Ring area, several residents, who contributed both to the sedentary interviews and the go-alongs, were rather uncertain about the design and function of the latest Bull Ring shopping centre.

There is a wider argument here about not just the removal or alteration of the built form but also the erasure of modernism. While and Pendlebury (2008) point out that it is important and relevant to understand such buildings not only in terms of their architectural qualities, but also in terms of the social intent that lay behind their production and why these values have been eroded. For some commentators there is a robust connection between the fluctuations of contemporary consumer society and the current landscapes of Western cities. A common characteristic of such spaces, Zukin (1998) argues, is an 'urban scanscape' perniciously surveyed by closed-circuit television and security forces, ever alert to those that might disrupt what Sibley (1995, page xi) calls 'the white-middle class family ambience' imbued in the design of 'new' public shopping spaces (see also, Raco, 2003).

Yet these are spaces that investors, developers and city officials seek to promote and to which many (though not all) contemporary consumers are attracted. These spaces are also controlled and managed in a variety of ways to displace certain behaviours, bodily practices and visual signs of 'disorder', eradicating or shielding them from view at certain places and times.

According to Oc and Tiesdell (1998), the city centre management of Coventry's precincts during the 1990s (operating under the control of the City Council, as the main landowner) saw the creation of bylaws as a means of curbing anti-social behaviour, and the installation of CCTV provided a solution to reduce incidents of criminal activity (Figure 6.2).

In other cases, this exclusion may be less overt and (arguably) more insidious. It would be fallacious to suggest that there were no attempts to control and order shoppers' activities within Coventry's 1950s shopping precincts or in the 1960s modernist Bull Ring Shopping Centre. In several ways, both supported consumption activities and were marketed to attract young families (especially females). New technologies, such as heating and escalators (in the Bull Ring) and car parking, were important attractors (while they worked). However, the latest Bull Ring could be interpreted as a highly regulated space, dedicated to a particular form of commerce/consumption for specific social-demographic groups, a point that was made by several respondents, including Raymond:

> Actually, I think in some ways the Bullring centre is more designed for youngsters. . . . Some of these Costa Coffee or Costa Lotto or whatever they're called . . . I think they're designed for the youngsters.
> (Raymond M., born 1938, interviewed 2008)[3]

While it is arguably a much nicer place to visit, particularly if you do happen to be the targeted middle-class consumer, the values of inclusiveness, which woven into the design of the modernist Bull Ring, have been watered down. Social inclusivity formed part of the designs of Coventry's Upper and Lower Precincts and Birmingham's old Bull Ring Shopping Centre. The Bull Ring Shopping Centre included upscale retail alongside a new outdoor

Figure 6.2 Bylaws and other measures to curb anti-social behaviour in the Bullring (left) and the Lower Precinct (right)

Source: Authors' own collection

market – to replicate the 'traditional' Bull Ring markets – that sat at the heart of the development. These sentiments were emphasised in the promotional literature that accompanied the opening of the Bull Ring Shopping Centre in 1964 that promised 'retail markets, shops, department stores, restaurants, public houses, banks, bus station, car park, baby crèche and offices' (Laing Development Co. 1964, page unknown). Paul, for example, remembered with some affection how these ideals worked in practice:

> I think erm generally Birmingham you know, everything seemed so much better [during the 1960s]. The whole era of the '60s, though, you know just the right time, everyone was full of optimism . . . [and in relation to the Bull Ring Shopping Centre]. I mean we got everything we wanted. The main thing was that we were interested in the clothes shops and the fashion shops . . . and the pubs [especially the Matador and the Toreador public houses].
>
> (Paul, born 1946, interviewed 2007)

It is tempting, therefore, to argue that the latest Bull Ring represents a highly regulated quasi-public space of 'enclosure', where the perspectives of different groups, including the elderly, are 'managed' or ignored. However, these narratives also demonstrate how respondents' memories can 'evade' official attempts to alter the 'bricks and mortar' of certain sites. There was also a feeling amongst some respondents that certain structures, perhaps, hold a certain enduring and even iconic charm. Most notably, James Roberts's Rotunda and Smallbrook Ringway were mentioned in this regard:

> I liked the Rotunda I thought it was wonderful. . . . At one stage they were going to pull down the building, and it is an icon of Birmingham now isn't it?
>
> (Kathleen, born 1950, go-along)

Resistance and change

It could also be argued that, as in other towns and cities such as Plymouth and Sheffield, the material legacy of post-war modernist redevelopment has been an under-appreciated aspect of Birmingham's and Coventry's heritage and design identity. Alongside other cities, the 'ugliness' of the post-war legacy is often criticised in popular, academic and professional representations. National opinions regarding the state of Coventry's built form tend re-affirm this sense of bleakness. For example, writing in *The Guardian*, Dyckhoff commented disparagingly that Coventry lacks a sense of identity 'no matter how much cosmetic surgery they give it' (*The Guardian*, 11 Mar, 2006).

To some extent, these attitudes have infused the current (and proposed) local authority policy landscape. Yet, there is some lukewarm recognition in Birmingham's *Big City Plan* that certain, though unspecified, '20th century buildings' in the city centre 'are often under-appreciated' (Birmingham City Council, 2011, page 35). The message contained in Coventry City Council's 'Core Strategy' submitted to the Planning Inspectorate in October 2012 is far more positive towards 'managing' the city's postwar legacy:

> All [future planning] proposals should aim to sustain and reinforce special character and conserve . . . Coventry's groundbreaking post-war reconstruction including its town plan, built form, art works and public spaces.
>
> (Coventry City Council, 2012, page 110)

Several respondents also commented of 'coming to terms' with the city centre, at least since the early-to-mid 1960s, being in a 'state of flux'. For Beryl, alterations to the city's built form cannot be divorced from the city's broader historical narrative; feelings of upheaval, change and 'progress' are intertwined with people's ongoing connection with the city:

> The city itself has always changed; from the estates – the [Eighteenth-century] Gooch, Jennens, etc., and they [the City Council] sold land for housing for factories [in the nineteenth century] and decisions were made. . . . Other historical events intervene – World War One and World War Two, etc. . . . So, the plans alter and change [because of these factors], or they become too expensive and they get shelved.
>
> (Beryl, born 1942, go-along)

Beryl was also sensitive to the break in historical continuity associated with the demolition of the city's post-war buildings. She argued that 'if you pull it all down, you lose a sense of continuity in history and . . . you have to [try and] keep the history of a place' (Beryl, born 1942, go-along).

Crucially, unlike Birmingham's Bull Ring, Coventry has yet to receive such a nationally significant 'flagship' retail project. As one of the key recommendations coming from Nathaniel Lichfield and Partners (2008, page 80) report on Coventry's retail offer suggests: 'Coventry needs to compete more effectively with large regional/sub-regional centres, such as Birmingham and Leicester and also other competing centres such as Solihull, Leamington Spa, Rugby and Nuneaton' (Clayton and Sivaev, 2013).[4] In Coventry, the well-intentioned, though 'limited', attempt to push through a coherent and consistent post-war architectural vision has, according to some practice and popular, academic and practice accounts, been at least partly responsible for the downturn in the city's economic vitality and reduction in the vibrancy of the city centre. Yet for others, especially those respondents who were in their late teens or early-to-mid-twenties during the rebuilding reported positive recollections about the post-war rebuilding. For example, some

Coventry respondents would often speak fondly of their feelings of 'adventure' (Dave, born 1941, go-along), curiosity and 'excitement' (John, born 1937, go-along) of having a 'new' city centre replete with modern facilities and new social spaces free from the danger of vehicular traffic. In this sense, the respondents' views chimed with Logie's (1962) assessment that the new city centre represented a national and international symbol of post-war rebuilding. Jean, for example, described the sense of enthusiasm of the Gibson era of reconstruction:

> There was nothing else like it wherever you went in other parts of the country, you know, you didn't see this kind of thing! It was so modern, and so nice for its time; it was pleasant, there were places you could sit, there was the trees, there was the flowers, there was the shops and like I say there was no traffic, it was great [. . .] you always felt that it was something for the future.
>
> (Jean, born 1936, interview)

Some Coventry residents spoke with great pride about how the overall design and preservation of the reconstruction-era architecture and public art has 'lasted the test of time' (Dave, born 1941, go-along). There was also a degree of enduring enthusiasm for the post-war public art. Respondents on the Coventry go-alongs were certainly drawn – and felt a certain affinity towards – the ceremonial Levelling Stone and its phoenix insignia (sculpted by Trevor Tennant and laid in 1946 and Listed in 2018) and the equestrian statue of Lady Godiva (1944, William Reid Dick, sculptor, Listed grade II*). The plinth is reminiscent of the work of Sir Edwin Lutyens, who died in 1944, but with whom Dick had worked. (Figure 6.3). Several respondents also favourably mentioned the tile panels by Gordon Cullen situated in the Lower Precinct (Figure 6.4). Less well appreciated was the Lady Godiva and Peeping Tom clock (designed by Trevor Tennant in 1953); mixed responses such as, 'oh, that awful thing' (Philip, go-along), were common during the go-alongs.[5] Interestingly, Birmingham had far less reconstruction-related public art.

Despite retaining several of the central elements of the Gibson plan, including the Upper and Lower Precincts, the circular market and the Broadgate buildings recent proposals for city-centre improvement have not evinced such positive reactions from residents and professionals alike in professional journals or newspapers (see, for example, A. Jones, 2011). The Friargate scheme (near to the city's railway station), comprising offices, hotels, bars, restaurants and (relocated) Council premises, is one development specifically designed to stimulate the local economy (*Coventry Evening Telegraph*, 21 May, 2013a) and increase Coventry's attractiveness to city centre users and investors as a location for knowledge intensive service businesses. The scheme aimed to radically alter junction six of Coventry's inner ring road, focussing 'pressure on the ring road, partly as result of additional traffic'

Figure 6.3 Lady Godiva
Source: Authors' own collection

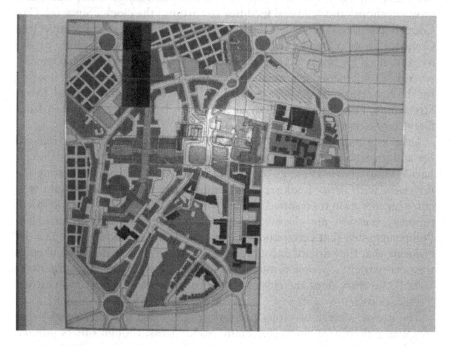

Figure 6.4 Lower Precinct Gordon Cullen tile mural
Source: Authors' own collection

generated 'but also because of the reprioritisation of road space required to enable high quality pedestrian and public transport accessibility' (Coventry City Council, 2012, page 94).

Unlike the removal of large sections of Birmingham's inner ring road, Coventry City Council has called for its inner ring road to be viewed as 'a major asset within the highway network which must also be protected'. In this sense, it 'enables the quality of the city centre environment to be maintained and prioritised for economic, retail and leisure activity' (Coventry City Council, 2012, page 88). For John H., the Broadgate area has ensured that vehicular traffic is now encouraged to use the inner ring road (rather than coming into the city centre):

> The Coventry Civic Society had a debate [about the ring road] and [one of the ideas was to] reduce the short stretch of elevated sections – and bring it down to grade, but with [the] pedestrianisation of Broadgate there is more of an emphasis on the using the ring road, but there was a view on making use of the road as a local road.
>
> (John, H. born 1931, interviewed 2012)

The City Council has sought to address the city's economic decline and improve its shopping offer from being the 'country's 13th largest city [but] with the 47th largest shopping centre' (*Coventry Evening Telegraph*, 21 May, 2013a). The Council is therefore keen to encourage 'vital, viable and growing shops' and 'cultural facilities . . . around the clock'. Since the City Council commissioned Nathaniel Lichfield and Partners to conduct the Coventry Shopping and Centres Study (in 2006) to assess the city's 'capacity' and 'need for future floorspace' (Coventry City Council, 2012, page 75), several developments have been approved, including, perhaps most notably for the city centre, the Bishopgate scheme (between Bishop Street and Tower Street) and Friargate (spanning junction six of the inner ring road from Greyfriars Green to the railway station).

Other respondents suggested that 'officials' should *not* continue to remember and (over) celebrate the post-war development, as it 'really holds the city back' (Philip, born 1938, go-along). This was a view espoused during the interview with Brian Redknap, who pointed out that the city should welcome regeneration to 'respond positively to the impact of the [late-twentieth century] recession' and the more recent (post-2008) economic decline. Here Philip suggested that certain people are keen to 'forget the (recent) past' by ensuring that they do not encounter objects, places or even people linked to the post-war city; 'people now shop at [nearby] Leamington' and choose to 'avoid' Coventry until the city's post-war environment is altered to attract 'better shops'.

The greater choice of shops and services provided by other nearby centres was a dominant theme for many respondents, though others pointed out that they might be attracted 'back' to shop in the city if there was an improved retail offer. Philip argued that the quality of shopping experience afforded by the range of shops located in Coventry's Upper and Lower precincts has steadily diminished over time meaning that, for him, the city centre does not hold much appeal when compared to other, more 'pleasant', shopping environments like nearby Birmingham:

> I don't come into Coventry very much anymore. . . . I've given up a little bit – They've wasted money left/right and centre. . . . You've got to get some decent chains and independents in, but . . . you'll never get Coventry like it was because people don't live here anymore. I've been to Plymouth and Birmingham and though there was a lot of argument about the design, but they knew what they were doing and the city [authorities are] good at promoting themselves.
>
> (Philip, born 1938, go-along)

John H., for example, pointed to what he identified as a general sense of ambivalence particularly amongst older residents (and groups such as the Civic Society) towards retaining remnants of the post-war architecture: he suggested that some people would welcome significant alterations. He also spoke of a certain uncomfortable 'love-hate' emotional attachment people seem to have with the Upper and Lower Precincts:

> I belong to the Civic Society and for the last 30 years and it has always been a contentious question whether [people] like it or don't like it . . . was it worth the money? These were questions being asked – [at] least in the thinking classes – [but] most people [were] just worried about getting on and getting by. . . . Coventry people have an innate emotional attachment, but they can't [help themselves from] criticising it – self-critical, really. It's a funny old place, but I love it. It's a love-hate relationship [especially] with the older people. Newer people have chosen to come here, haven't they? So, they have a slightly different perspective, perhaps.
>
> (John H., born 1931, interviewed 2012)

There were also strong feelings among many that recent changes have been less than sympathetically handled. Others acerbically commented on the 'awkward' design of the escalator that leads to the West Orchards shopping complex from the Upper Precinct (Figure 6.5).[6] Several respondents mentioned how the view to the Cathedral – a key aspect of Gibson's earlier ideas – was later blocked by the Cathedral Lanes development. Special criticism was reserved for the detrimental impact the way in which the tower blocks compete with the three spires – especially the views to the Cathedral.

Since the late 1980s, however, there have been countervailing forces which seek to preserve elements of this 'unruly' past (Ashworth and Tunbridge, 1996). Post-war architecture also has come under the ambit of state

Figure 6.5 Upper Precinct complete with additional escalator and the Cathedral
            Lanes development
Source: Authors' own collection

protection, with over 400 post-war buildings or structures being granted
listed status by the UK government (see https://historicengland.org.uk/). In a
Scottish context, for example, Gillon and McDowell (2012, page 5) continue
the theme of 'progressive' listing, suggesting that listing should be an 'ongoing process that recognises changing levels of knowledge and that every gen-
eration has its own view of what comprises its heritage'. Aspects of 1950s
and 1960s architecture have also become fashionable for urban design and
style elites. Conservation-led re-commodification of post-war buildings has
also centred on individual 'iconic' structures that can be woven into the
wider narratives of consumer-led regeneration. There have also been pockets of noticeable, albeit local, resistance to the unceremoniously 'disposing'
of elements of Birmingham's post-war city. As with other reconstructed cities (e.g. Plymouth, [see Gould, 2000]) some of the city's leading post-war
architects have been an important part of a broader lobbying movement for
the retention of post-war architecture (see www.thebirminghampress.com).
When interviewed, John Madin stridently argued that it would be the definitive act of urban regeneration to take the existing library and bring it back
to life, but for a cost not unlike that required for the new building.

The Twentieth Century Society has had a central role in promoting the
need to protect buildings, like Madin's Central Library. It is also significant,
perhaps, that recent architectural accounts, including the updated volume
in the acclaimed 'Buildings of England' series (Foster, 2005, 2009; Pickford
and Pevsner, 2016), have constructed a mostly compassionate perspective
of the 1950s and 1960s built legacy. And Madin's own contributions have
been subject to critical reappraisal (Clawley, 2011) even before his death

in January 2012 (see also Beanland, 2013). For example, the Birmingham Architecture Festival 2013 (held between May and June 2013) presented, among other things, a photographic exhibition of the Central Library by architectural photographer Richard Southall, intended to provide a definitive architectural record of a controversial and threatened structure (see also C. Madin, 2011). In Birmingham, the City Council has also been keen to locally list several 1960s buildings, including Seifert's Grade II listed Alpha Tower, Broad Street and championing 1950s and 1960s built heritage (see Birmingham City Council, 1999), while the city core is also covered by a conservation area (designated in 1971 and extended in 1985) (Figure 6.6).[7]

Certain 'iconic' structures, such as the Rotunda (Figure 6.7) and Mailbox, arguably manage to sit rather comfortably beside the more recently constructed buildings, and this example fits comfortably within the broader narrative of post-industrial urban renaissance. In the Rotunda's case, though

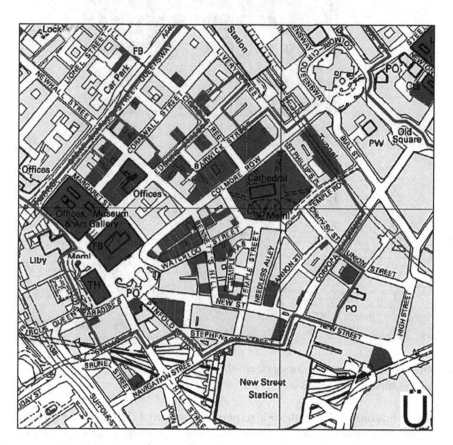

Figure 6.6 Statutorily Listed Buildings in central Birmingham. The map also shows the extent of the city's conservation area

Source: Crown copyright reserved

Figure 6.7 The Rotunda during refurbishment (2006)
Source: Authors' own collection

listing has been suitably flexible to permit substantial change to the physical fabric, despite public and professional doubts regarding whether the alteration to the original design has adversely affected the architectural merit of the building. James Roberts himself supported the radical refurbishment. The lure of having large, flexible floorplates for residential or office stock

is also blended with a particular urban 'coolness', as reflected in the transformation of buildings like the Mailbox. The Mailbox, formerly one of the
largest Royal Mail sorting offices, dating from the 1960s, likewise became
redundant in the late 1990s. Stripped to the steel frame the building now
accommodates high-quality retail space, offices and restaurants, with the
BBC a major tenant, having moved from their now-demolished Pebble Mill
studios elsewhere in the city.

From a contemporary construction perspective, some post-war buildings are now idea. Once acquired, they can offer flexibility because many
are steel-framed structures and relatively easy to sub-divide. It was for
these reasons (and others) that, in the 1990s, the developer Urban Splash[8]
was vigorously purchasing 1960s Birmingham's stock of 'out-moded'
office and industrial buildings. While features of the original design intention have been maintained, the Urban Splash approach has involved an
external renovation: recladding and/or repainting dirty concrete, together
with other design modifications such as new entrances, but where possible window and other original features are retained for economic as
much as design reasons (While and Short, 2011). John Madin was also
keen to emphasise that his original idea was to clad the Central Library
in Portland stone or travertine marble, set in landscaped gardens replete
with fountains and waterfalls. Instead, the City Council adopted a cheaper
solution using pre-cast concrete with a stone aggregate. This led to later
criticisms that the library was a 'concrete monstrosity' (see HRH Prince of
Wales, 1989, page 32) (Figure 6.8).

Figure 6.8 Demolition of Madin's Central Library (2016)
Source: Authors' own collection

Curiously, perhaps, the Council also cited the failure of some of the concrete panels in 1999 as a reason to demolish the library and pass the site to a commercial developer. Madin suggested that the decision by the City Council in the late 1980s to enclose and privatise the central atrium of the Central Library also proved to be problematic:

> I designed the library as a civic square with fountains and waterfalls; this [has] been closed off. [But] the whole civic square has [since] been filled with fast food, in the very heart of the civic centre of Birmingham!

(Madin, interview)

These ideas regarding contestation and change have been brought into sharper focus given the decision in 2009 by the then UK government to refuse English Heritage's second recommendation that the library should be listed. In December 2012, proposals to demolish the 1974 built Birmingham Central Library and build new offices, shops and public squares were unanimously backed by the council's planning committee (*Birmingham Post*, 2 Dec, 2012a). On reflection, such an acerbic assessment is unkind: when discussing the NatWest Bank, Madin pointed out that the he was working within the parameters of the brief as set out by his client. He suggested that 'they required . . . [a] sort of regional centre offices, that's why it was such a tall tower'. Madin stressed that: 'I just tried to meet their requirements and at the same time do it in such a way, so it wasn't out of scale with the existing buildings in Colmore Row'.

Elsewhere within the city centre other post-war buildings have received more recent praise for showing hallmarks of conservative late modernism. Within the city centre, the stone-fronted 23–24 Bennett's Hill, by the Stafford-based architect E Bower Norris (1961), and Grosvenor House on New Street by Cotton, Ballard and Blow (built 1951–1953) are notable in design terms; differing from the familiar modernism of so much post-war reconstruction (Figure 6.9). Harwood (2003, page 210) suggests that Grosvenor House provides an example of 'rare wit and integrity in the design of speculative office buildings'. It was Listed Grade II in 1999 as 'an unusually cohesive design . . . an imaginative and richly detailed example of a 1950s speculative office building erected before the removal of wartime restrictions on construction . . . in 1954' (https://historicengland.org.uk/listing/the-list/list-entry/1245547).

Certain developers have warmly embraced the opportunity to redevelop some prominent post-war buildings in the UK. While (2006), for example, demonstrates that city authorities in places like Birmingham and Coventry, have largely encouraged development that significantly removes vestiges of prominent 1950s and 1960s reconstruction projects. In central Birmingham, the 'reconstruction identity' of some buildings is now obscured by recladding.

Figure 6.9 Grosvenor House, New Street
Source: Authors' own collection

Memory and the senses

There is considerable literature on the design of post-industrial change
and the influence of the *visual* form of regeneration projects in Western
cities. Some commentators argue that the emphasis placed on promotion
and 'branding' of Western cities by city officials inescapably results in an
abundance of visually similar 'clean, safe and predictable' (Zukin, 2010,
page 128) retail spaces – essentially 'cloned, banal, branded landscapes'
made up of 'central city malls and [other] regenerated spaces' (Tallon,
2010, page 20). Such spaces arguably do little to encourage a sense of social
vibrancy and 'sense of place', or to engage people as they move through
these spaces. Recent measures to address these broad concerns have been
enshrined in official English planning that provide 'a diverse retail offer
and which reflect the individuality of town centres' (DCLG, 2012) and to
create 'lively, dynamic and exciting social places' (Portas, 2011, page 3),

responding to the 'scare' of the 'clone town' (New Economics Foundation, 2010). This section reviews respondents' feelings about recent efforts to provide a visually stimulating and socially vibrant townscape and brings into view the importance of personal memories in shaping people's current understanding and experiences of Birmingham and Coventry, in the context of their past experiences.

The desire to generate a more visually enticing and locally distinctive retail offer is of some considerable concern for city officials in Coventry and Birmingham. According to Coventry City Council, 'we can't let the people of Coventry (and the developers) down by allowing identikit shopping malls to spring up to replace the tired shops that have grown up since the city centre was rebuilt after the war' (cited in *Birmingham Post*, 27 Feb, 2012b). The message of improving the visual attractiveness of the city comes through the City Council's Core Strategy (Coventry City Council, 2012, page 79), which states that the 'the city centre' will, by 2028, 'continue to be developed to make it a showcase for the city'. Most of the Birmingham respondents conveyed a general sense that they were not necessarily attracted by the attempts to 'beautify' the latest Bull Ring shopping centre (cf. Parker and Long, 2004).

Moreover, they tended to comment critically on how the variety of post-war Birmingham has been lost in recent years as smaller retailers are being displaced by larger national and international multiples. This was a view echoed in one letter to the local newspaper in 2003 which stated that the 'Bullring will suck the life out of smaller shops . . . all roads lead to the new complex without a thought for the effect on small independent stores' (*Birmingham Evening Mail*, 30 Aug, 2003). Although this is symptomatic of a much broader trend experienced across British high streets in recent years, one respondent – Maureen – complained that such a tendency has had a 'homogenising' effect on the city centre. Comments relating to the design of Selfridges ranged from 'horrible . . . it is completely out of character' (Malcolm, born 1947, go-along), to 'wow, it's fantastic' (Paul, born, 1946, interviewed 2007). There was also a general sense amongst the Birmingham respondents that shopping trips to the centre of town, including the Bull Ring, were largely reserved for 'special items', with several people preferring to shop locally for their everyday needs. A similar narrative came forward from Raymond during the Coventry go-alongs: here he complained that 'you get the same shops in every town now', before going on to recall how his father used to be attracted to the variety afforded by the types of goods sold in the Barracks Market before the war (Raymond, born 1937, go-along).

There are also attendant assumptions permeating much of the literature on urban design that the design (or even the redesign) of regenerated spaces has an important influence upon, though not entirely deterministically, 'the probabilities of certain actions or behaviours occurring' (Carmona *et al.*,

2009, page 134) and their current experiences of the built environment. Rather less academic (and practical) attention has been devoted to the way in which 'everyday' personal memories shape *different* kinds of sensorial urban experiences. In this sense, the go-alongs prompted rich sensorial recollections of how these city centres were once structured and used and, importantly, recalling these memories was part of how these respondents currently interpret the city centres. The present is compared to the past in a variety of ways and frequencies.

Official narratives can induce emotional reactions, of course. Certainly, the events of the Second World War have become part of the cultural vocabulary of the city: one that remains an important symbol in the human demand for theodicy – the need to commemorate the injustices of war – not just in Coventry but also elsewhere. Instead of being encased in 'spatially fixed' nostalgia (e.g. Harvey, 1989), several respondents spoke of how they continue to connect with Coventry's role as an officially recognised city of peace and post-war reconciliation. As detailed below, the go-alongs helped to bring out individuals' largely spontaneous recollections of both Coventry and other war-ravaged and reconstructed cities in the UK and Europe. In this sense, urban materiality 'escapes' its immediate spatial context as respondents walked around and talked about their city. Haughton and Martin (2014, pages 210–211) suggest that even when 'certain aspects of the past appear not to be salient any more' they have a habit of returning [leaving] landscapes [that] can be perfectly habitable, but at any one moment [citizens] can be reminded of a landscape that is difficult to traverse and navigate.

Perhaps more importantly, Coventry's international significance also lies in its designation as a city of peace and reconciliation centred on the Cathedral of St Michael, the University's Centre for the Study of Forgiveness and Reconciliation and the city's 'special' relationship with other bombed cities, including Dresden. Goebel (2011, page 167) argues that Coventry became an international symbol around which Coventrians became 'spun [into] a worldwide web of commemorative partnerships'. During the early 1960s, Coventry became entwined with 23 other cities across the world.

Arguably, it is Coventry's connection to Dresden that is perhaps the most enduring; with both cities sending and receiving official and semi-official delegations, holding exhibitions and formalising a friendship pact in 1963–1964' (Goebel, 2011). Despite the Bishop of Coventry's announcement during the commemoration of the 66th anniversary of the destruction of Dresden in 2011 of how 'Dresden and Coventry are bound together in a history of suffering and in a future of hope' (*Church of England*, 25 Jan, 2011, page 1), the link between the two cities has not always been so strong. By the mid-1960s, though, Coventry politicians distanced themselves from the idea of equal suffering: a Coventry councillor suggesting that 'the suffering of Coventry was not as great as the suffering that was in Dresden' (CCA, PA/732/1/18/3, May 1965). It is interesting to note this recognition – it is still not always admitted, in the authors' experiences of speaking to local groups in England! Basil Spence's design for the Cathedral, the 'centrepiece of the reconstruction' (Gold, 2007, page 79), continues to strike a popular 'chord' with locals and visitors and its materials – sandstone, Westmorland slate, copper roofing marble flooring – are broadly in-keeping with the architecture of Gibson's approach (Gould and Gould, 2016, page 124). While seated on a bench in the ruined shell of the medieval Cathedral – deliberately left as landmark and symbol of its own destruction and resurrection – Barry talked positively of the bond that existed between the two cities. However, he suggested, that his reminiscences of growing up in the Hillfields area of Coventry after the war were not necessarily comparable with the those living in war-ravaged Dresden:

> I went to Dresden recently and people in Dresden talked to me very nicely and they think Coventry had it as bad as they did, but I don't think we did at all. I don't remember going hungry because we always knew that time in the early-50s, I suppose, from about 1948 onwards . . . when we used to play in the streets, the women used to go out shopping and one woman used to take a turn and every night [to do the shopping], we seemed to have bread and jam and we just accepted it, really, we knew that we would get bread, butter and jam and I can't remember going without and hungry – not like those in Dresden. But they were talking to me like I was their next-door neighbour – there was a kind of bond.
>
> (Barry, born 1937, interviewed 2012)

There are also less prominent visual cues of the city's recent past. For example, during one Coventry go-along, while crossing Corporation Street from the Collegiate and Parish Church of Saint John the Baptist, Spon Street, Raymond noticed an almost concealed plaque commemorating the 'friendship link' shared between Coventry and Lidice, a village north-west of Prague, partially destroyed in 1942 by German troops. This is situated on a wall as a reminder, a spatial prompt that 'calls' to passers-by – although it is hardly a prominent feature. Although Raymond recalled a painful recollection, he also talked through the importance of remembering such an event and hinted at the importance of communicating messages to current and future generations:

> That circle over there is about Lidice Place in the war. It's about [Reinhard] Heydrich – the Nazi commander in Czechoslovakia, I think, he was ambushed by partisans there. And there were some terrible stories [of how] the Nazis razed the village where they [partisans] came

from. . . . We forget too easily the horrors of all war; we forget too easily what happened in Coventry and elsewhere.

(Raymond, born 1937, go-along)

However, in his discussion of Coventry's relationship with other bombed European cities, Goebel (2011, page 169) cautions against the over-emphasis of the shared wartime and post-war experiences included in the physical landscape or circulated via professional publications, art installations, films and exhibitions. Coventry's pioneering reconstruction architecture, so widely admired in the post-war years, has arguably managed to overshadow earlier examples of heritage and 'somehow erase the collective memories of place, stretching back almost a millennium' (Walters, 2014, page 8). For instance, while walking through Lady Herbert Gardens, Raymond also spoke of the way in which the post-war legacy is perhaps over-played at the expense of other historical periods:

Here we are in Lady Herbert Gardens; it's beautiful . . . it was run down for a time [but] I used to walk in there with my parents or come on my own. It's lovely, the only real surviving gate in Coventry. But I'd regularly walk through there it was beautiful and of course, when you think there was quite a small perimeter of the [medieval] wall; but it's a bit overlooked, really, with all the focus on the blitz and the rebuilding.

(Raymond, born 1937, go-along)

It is significant to note that different 'sensory' qualities came through in relation to other reconstructed cities, too. For example, emerging from Paradise Forum beneath Birmingham's Central Library, and passing into Chamberlain Square, Maggie took a moment to stop and glance at the Town Hall and the Council House, which led her to draw comparisons with post-war Berlin. Here she contrasted devastated Berlin and the sharp 'broken tooth' like appearance of bombed-out buildings in Potsdam Platz with the undamaged buildings encircling Birmingham's Chamberlain Square:

I tell you what, when I was 20, we went on a tour of Europe with an American ballet company and of course, I was only a baby, a little thing during the war, so I didn't really know about it and my mum and dad didn't really tell me much, either. You know bombs, dropping, men dying, you know, that sort of thing. And . . . I went to Berlin and it was when they had Russian, French, British and French zones, and we were on Potsdam Platz which stuck out like a broken tooth, and I didn't know that Potsdam Platz was the centre, like the middle of London, Trafalgar Square, and the whatsit building . . . the Reichstag was just a shell. Near it was the railway station and I couldn't get my head around it, and now it is something quite remarkable. [It is] fascinating how

things changed and it's nothing short of a miracle that this lot survived
. . . you know.

(Maggie, born 1937, go-along)

As the above narratives indicate, there is an 'agency' to urban materiality –
buildings, objects and infrastructure –, which can influence people's experiencing, of the present. This narrative also problematises the notion of
identifying or foretelling which material object can encourage past recollections: there is no straightforward way of predicting which [material]
object will hold cues to the past (Muzaini, 2015), thus making it difficult for
individuals to adopt effective strategies of remembering or forgetting. Furthermore, recollections are also heightened with the multi-sensual engagement with places, people and objects. For example, during one go-along,
Val, while walking behind John Madin's Central Library, was struck by the
sights and sounds of passing traffic along Great Charles Street (part of the
city's inner ring road development). Her narrative also reveals how troublesome memories can be involuntarily triggered by absent sites.

Rather than speak about the immediate environment of the Central
Library or the ring road, Val proceeded to recall the time when she was
working at a school in the 1970s and her memories of the disruption/dislo-
cation caused to local schoolchildren and their families by the modernisation of the city. The sounds and sights of vehicles flowing along the road
affected Val, thus also demonstrating how sensorial memories can affect
individuals' present experiencing:

> Yes, this created a lot of upheaval. . . . I went to [teach at] a school on
> Castle Vale . . . a lot of children said how unhappy they were because
> their families weren't there . . . they moved them out [by] getting rid of
> the back-to-backs. I remember the children crying, saying: 'my mum's so
> unhappy because mum can't see grandma'. It really upset them because
> the community had been broken up. It really upset them having lived in
> a close community.
>
> (Val, born 1932, go-along)

During the go-alongs, while looking at St Martin's Church and at the new
Bullring, Kathleen's recollection hints at the blending between the auditory,
ocular and olfactory senses, which act as 'important frames of remembering' (Low, 2013, page 704). Kathleen proceeded to describe her once regular
route from the Rotunda, via a subway, into the Digbeth area of the city during the 1970s as being 'horrible', 'noisy', 'dark' and 'smelly'. While there are
no specific visual cues left of the 1964 Bull Ring shopping centre, her recollection also connects to the point about how emotional, bodily and mental
'wounds' can sometimes spark adverse memories of the past:

> I worked for Lloyds Bank; they had a branch in the early 1970s at the
> side of the [old] Bull Ring [Shopping Centre] past the Rotunda and

it was really horrible to go down there! I didn't like it. You had to go through underpasses – there [points] little . . . passages, and it was always crowded. . . . I had to walk down the subway, through the open market and it was literally underneath the multi-storey car-park, but it was horrible, noisy, smelly, [and] so dark! No, I've had quite enough of Birmingham over the years!

(Kathleen, born 1950, go-along)

This go-along therefore reminded Kathleen about how she had 'quite enough of Birmingham . . . over the years', before going on to talk about how she now chooses to limit her journeys into the city, as a way – at least in part – of restricting uninvited reminders of her walks around the reconstructed city during the 1970s. And while memories of the post-war landscape may be triggered through a blending of the various senses, these recollections are also rooted in an individual's consciousness, 'haunting' their existence, ready to 'burst through' everyday experiencing. In this sense, the mind and memory are not solely in the brain but are more profoundly dependent on body states too; circulating hormones, physiological process, the immune system, all interact reciprocally with the brain in ways that mean that brain and body – and hence mind and body – cannot easily be separated. Walking and interviewing *in situ*, open to environmental stimuli (and other sensory 'triggers' – different noises, smells, as well as the visual), was a particularly useful tool to unearth people's recollections. The go-alongs also helped respondents construct memories that were rather less about past events tied to the post-war building but were more about a materiality that is no longer present.

What was particularly evident through the go-along approach was how the immediate environment motivated people's remembrances. Walking also assisted with creating a focus on the specific details with ordinarily commonplace, 'lost from view' elements of the streetscape that can stimulate reactions from participants. The elicited narratives spoke of how the experience of the built environment in the present context is blanketed with recollections of how places were experienced in the past. During one go-along perambulation along an (altered) section of Birmingham's post-war inner ring road, a passing siren outside Moor Street Railway Station in Birmingham immediately struck Malcolm – who was studying at night school during the early 1970s. On hearing this, Malcolm turned to look up at the Rotunda and recollected the time when the Mulberry Bush Pub, situated at the base of the Rotunda (in New Street), was bombed, not from a Second World War German air raid, but by the IRA attack of November 1974 (where it is estimated that around 182 people were injured and 21 people died). In a sense, 'lost' sites, such as the 'Mulberry Bush' and the nearby 'Tavern in the Town' were 'seen' (and 'heard') through memories of what was once there but is no longer (see Adams and Larkham, 2016). For some, this presented a turning point in the building's fortunes. While standing at the base of the Rotunda, memories of this event seemed to be very strong:

> The bombs went off here [standing at base of Rotunda] – I was on the annual Christmas do at the [police] station I was working that night [when it went off]. We had thousands of police officers at the airports, but it happened here instead. There used to be a gully [down there – points] and that's where the bomb was – one of largest peace time atrocities, I think 23 died [*sic*] I think. I did go in there when Lloyds Bank were [*sic*] in the Rotunda and then office spaces and that. It is a landmark really against the attacks.
>
> (Henry, born 1941, go-along)

> We've walked past the 'Yard of Ale' which was bombed [then named the Tavern in the Town], it's not there anymore – that was terribly shocking to all of us because all of us young people used to meet in there. [It was] my wife's last Christmas at work [and she] was pregnant with our son there and we were supposed to go to the Bull Ring Shopping Centre for dinner . . . but we never went in the end. It was never quite the same. The Rotunda itself was fantastic . . . but it survived two or three attempts and it became a symbol sort of Birmingham stubbornness and resilience – we won't give in!
>
> (Steven, born 1949, go-along)

Conclusion

The built form of both Coventry and Birmingham has changed dramatically during the past 50 years. Of concern perhaps, were respondents' views towards a perceived 'lack of' suitable retail spaces and major anchor stores or significant supermarkets within Coventry – for some people this issue, along with the city centre's perceived lack of social vibrancy, remains a long-standing and unresolved 'problem' that still needs to be addressed in any future regeneration proposals. Recent changes to the post-war buildings have not necessarily helped Coventry's cause in this regard. Both the City Tavern (built in 1958), a copper-clad façade between the Locarno Ballroom and the western range of shops, in Smithford Way and the Market Tavern (built in 1956), a self-supporting two-storey brick box, in Market Way, intended to create a sense of vitality in the Precinct during the evening, have subsequently been demolished and replaced with shops. These changes rather crudely ape the architectural style of their near neighbours (Gould and Gould, 2016). Coventry Precinct, like Exeter, was always intended to be a pedestrian-friendly design, and although it is of a pedestrian scale, there have been some significant alterations in the distances between the shops. More recently, the landscape and paving have also been altered; nowhere more so than in the Upper Precinct and in Shelton Square where the access to the upper levels has been modified.

Of course, anxieties were certainly felt by Coventry respondents over the design legacy associated with much of the built form (especially the precincts) and how this impacted people's contemporary shopping excursions and experiences, while concerns were also raised over the re-packaging of the Spon Street area during the late 1960s and early 1970s. Despite this, however, feelings towards the post-war townscape were rather mixed. For example, although criticised as being architecturally unremarkable, the original use of Blockley brick with contrasting stone detailing does provides certain enduring harmony to Coventry's shopping precinct and the civic quarter – a point on which most respondents agreed. The introduction of the terminal towers in the Precinct during the mid-1950s (when Arthur Ling assumed the role of City Architect) used different materials, including concrete, to those favoured by Gibson. Hillman House (built in 1962 and designed by Arthur Swift and Partners) and the much later Coventry Point (built in 1978; designed by the John Madin Design Group) are examples in this regard. Some Coventrians felt affinity and attachment towards aspects of public artwork, including, though not exclusively, the Lady Godiva statue, the Levelling Stone and the Gordon Cullen murals.

The lack of an overall driving idea that was arguably a limiting factor for Birmingham's post-war architecture. There was no obvious architectural arrangement, and there was little uniformity in elevational rhythms, materials or styles. Although quite different in detail from Coventry, Plymouth or parts of Exeter, central Birmingham's built form might be characterised as being a confusingly diverse set of masses, elevations and designs and distinct façades resulting in an under-whelming collection of post-war buildings (cf. Foster, 2005; Gold, 2007). Furthermore, Birmingham had no grand public spaces like Coventry Precinct or Civic Centre, Exeter's High Street or to Plymouth's Armada Way, and there is a rather limited idea of being part of a grand plan as at Plymouth. Chamberlain Square was a popular space between the Art Gallery and Central Library, but hardly 'grand' – nor were Manzoni Gardens. Most of the Birmingham respondents recruited for the sedentary and walking interviews reported that several post-war buildings were designed and built in a very 'blocky', 'childlike' way, and some people also lamented what they perceived to be an over-reliance on the use of concrete as a building material. For many of the Birmingham respondents, therefore, the passing of time has done little to change their perspective of the city centre's built form. However, they were rather less specific over precisely which buildings fell into such a category. Buildings such as Madin's Central Library were certainly mentioned by respondents in this context; this is entirely understandable given the level of local (and national) media interest in the future of the site at the time of the interviews. Other buildings, such as Roberts's cylindrical design for the Rotunda and the narrowness of the sinuous office block aligned with the first section of the inner ring road at Smallbrook Ringway and with its shop units addressing the road were also singled out for particular praise. While the Rotunda has been Listed, the Smallbrook Ringway block has been granted a Certificate of Immunity from Listing.

Although it is difficult to provide a coherent narrative for the centre of Birmingham – especially when, along with Roberts and Madin, a range of architects was active in bringing forward designs for city centre buildings, most people rather vaguely suggested that 'something' needed to be done with certain aspects of the city's post-war legacy. The contemporary desire or need to attract private finance through new development projects is enmeshed in policy discourse with rather lukewarm recognition of the importance of post-war structures – even in the face of some stiff, well-marshalled and in the case of Madin's Central Library, local resistance. Several people also applauded the ways in which certain surviving post-war buildings – especially the Rotunda – have been successfully retained and re-worked into the contemporary townscape. However, these responses were at least partially offset by a tacit concern that some of the social values that were broadly enshrined in the City Council's post-war ambitions of moving towards a new city have changed significantly. While those ideas were embedded in the design (and reality) of the Bull Ring Shopping Centre, for example, they have been diluted in the recent highly visual, stylised and consumer-led regeneration initiatives.

The story is slightly different for Coventry. The Cathedral Lanes shopping centre on Broadgate has both compromised the open-ness of the square and blocked the visual axis between the Cathedral spire and the Precinct. The West Orchards shopping centre has adversely affected the original design intention of the precinct and the north end of Smithford Way. There have been some rather mixed reactions to recent attempts to remove selected elements of the post-war townscape. Respondents recorded more of a sense of ambivalence towards retaining the city's post-war legacy, with some noting that the architecture has really lasted the test of time despite piecemeal changes. Whereas others feel that, a more radical approach is required to attract flagship stores and thus arrest the economic decline of the city, making Coventry more competitive at a sub-regional and national scale. These views are at odds with the earlier decision to nationally list the Rotunda cafe in the Lower Precinct, a decision which, as While (2006) notes, has effectively blocked wholesale changes of the precincts and the most recent (2018) Listing decisions.

The buildings of the shopping centres of Coventry, Plymouth, Exeter and Bristol's Broadmead were intended to be altered over time. The extent to which they would have to adapt to reflect changing demands of the retail industry could not have been foreseen in detail by the original developers or architects. But in only a few cases have the buildings changed dramatically away from their original design; and only in Exeter and Broadmead and The Moor, Sheffield, have there been major demolitions. Un-planned central Birmingham has also suffered in this respect. However, one could argue that the longer these buildings and infrastructure survive, the more likely they are to be appreciated – both professionally and publicly (cf. While and

Short, 2011); although, of course, small, incremental changes can also damage the architectural qualities significantly over a long enough time span. Alterations to the streetscape, shopfronts, fascias and windows all make a major difference – as, indeed, does neglect, especially of upper-floor façades and particularly of concrete structures.

Despite the recent attempts made to regenerate both Birmingham and Coventry – including, especially in Birmingham's case, design-consciousness to promote new development and create a socially vibrant city centre – the narratives collected during the round of go-alongs also spoke of how places are experienced in multi-sensorial ways and how memories of the built form are also important. These are arguably under-appreciated aspects in shaping and mediating people's current experiencing of places. This was more noticeable for those Birmingham respondents who could remember specific places within the city centre. This was certainly the case for some respondents who fondly recalled childhood memories of using the Bull Ring during the 1950s and reflecting on how the built form in this space has, effectively, undergone two significant transformations since the 1950s. For Coventry respondents, in contrast, the go-alongs produced some comparisons with other places – especially and understandably, Dresden and less obviously, perhaps, Lidice.

Notes

1 Roberts may have been referring to the Peter Jones store, by William Crabtree in 1936, though owned by the John Lewis Group.
2 Figures from the Census of Population (2011) suggest that Birmingham Metropolitan District is made up of 37% 0–24-year-old resident population – the highest proportion of young people in 'any other European city' (Birmingham City Council, 2011, page 8).
3 It is also acknowledged in a report by DTZ (2013, page 25) that, given that Birmingham's current shopping environment has a predominance of high-end retail outlets and 'a young focus', 'some lower-value retailing could be welcome in the [city] centre'.
4 Clayton and Sivaev (2013, page 35) recommend that the City South Scheme should help to 'increase the density amongst firms and other organisations' leading to 'a significant increase in innovation capacity in the area'.
5 See Nozlopy (2003) for a full description of public sculpture in Coventry.
6 Plans are in place to remove the escalator, as part of the Council's efforts to revitalise the main shopping area and improve the public ream (see: https://coventryobserver.co.uk/news/plans-to-remove-hated-precinct-escalator-move-forward/).
7 Note, however, that the western edge of the conservation area was carefully delineated such that Madin's Central Library lay just beyond it.
8 Urban Splash has specialised in refurbishing 'difficult' buildings. For example, after falling into disrepair, Sheffield's Park Hill council housing estate, built between 1957 and 1961, has since been renovated by Urban Splash. Park Hill was granted Grade II* listed building status in 1988 and the renovated show flats have embraced a 'cool', 'hipster' approach incorporating 'jaunty, funky, upbeat, witty interiors' and 'the estate's most famous graffito – "I love you will u marry me", sprayed on a high bridge – has been memorialised in neon' (Moore, cited in *The Guardian* 21 Aug, 2011). In 2013, Park Hill was shortlisted for the prestigious RIBA Stirling Prize.

7 Moving from vision to reality

Implementing the reconstruction

While many modernist-inspired professionals, from architects and survey-ors to planners, might have dreamt of assuming an omnipresent 'god-like' status charged with planning whole cities, it could be argued that the power of the modern state created the space for local post-war boosterist modern visions to be realised. In realising them, however, the state and opportunistic developers could be accused of producing what have subsequently been characterised as some decidedly dystopian modernist urban environments: Britain's municipal tower blocks, the United States' 'failed' housing 'projects',[1] France's *banlieus* and a litany of others (for an overview of these broad arguments, see Platt, 2015). *Tabula rasa*-style planning is underpinned with modernist assumptions about the production and consumption of the built environment and, increasingly, post-war buildings and infrastructure are condemned as irredeemable and demolished to make way for an arguably more fragmented vision of the future.

There is no denying that the complex process of urban planning, design and development has its shortcomings, its totalitarian compulsions and its failures. Contemporary professional and popular cultures are also suffused with images, discussions and representations of dystopia and urban decline, disease, crime, pollution, degradation and disorder. Much of the post-Second World War built form in certain Western cities can be read as being symbolic of this urban malaise, but it is something that can be changed: either demolished and replaced or reshaped and rebranded as 'cool' for post-industrial middle classes. It is also difficult to dismiss the argument that too many studies romanticise the lives, works, achievements and overall influence of eminent 'planners' in shaping past and present cities. It is also problematic to unquestioningly accept the value of bottom-up perspectives and realities of how urban space is made/remade. Not all 'planners' neatly fit the profile of omnipotent heroic utopian visionary; and not all consumers of city spaces feel the controlling forces of urban decision-makers. The late-twentieth-century trend of giving voice to alternative bottom-up perspectives persists. This offers an alluring counterpoint to the remoteness of grand theorising and/or a focus on the life and influence of the 'great planner' sweeping away all before him.

However, of course, there are important shades to this general line of argument; not all cities approach large-scale rebuilding in the same way, and the personalities and attitudes of those involved in the rebuilding differed. In a British context, Plymouth, for example, represents the culmination of Abercrombie's grand and formal plan (though even writing this common attribution downplays the input of J. Paton Watson, the City Engineer and Surveyor and first-named plan author), while Sharp's Exeter plan invoked a spirit of conservation while simultaneously seeking to incorporate new spaces and new civic vistas into elements of the surviving urban fabric. For Bristol, the Broadmead shopping centre was the consequence of development driven principally by commercial considerations. Plymouth was the most powerfully placed British city when we consider the issue of political governance: prominent actors such as Lord Astor, the wartime mayor of Plymouth and his wife, Nancy, who was MP for Plymouth Sutton from 1919–1945, were dominant and well-connected (Essex and Brayshay, 2013). Through their strong political ties with Lord Reith, the Minister of Works, they succeeded in appointing the much sought-after Patrick Abercrombie as architect-planner for Plymouth in 1942 (Hasegawa, 2013). Lord Astor enjoyed cross-party political support: from the right, from Sir Clifford Tozer, who became chairman in 1945 and from the political left, from Alderman Mason, the deputy chairman and also from Hubert Medland, the MP for Plymouth Drake, who had a strong personal relationship with Evelyn Sharp, Deputy Secretary at the Ministry of Town and Country Planning (later Dame Evelyn, a formidable and influential civil servant) (see Gould, 2000). Abercrombie, with his contribution to the Barlow Commission, he had direct connections to the wartime government. He was influential in both establishing the Ministry of Town and Country Planning in 1943 and the drafting of the 1944 Town and Country Planning Act, which set out the principles of the compulsory purchase of city centre land after the War (see, for example, Cherry, 1981).

In Abercrombie, for example, Plymouth had succeeded not only in having the most pragmatic architect-planner available to lead the process of reconstruction but also, he was someone who could appreciate – and successfully negotiate ways around – the complex legislation of post-war government in a way that benefitted the city (Gould, 2000). Even so, the Paton Watson-Abercrombie plan was not implemented in its original form, and some elements of its layout were changed considerably. And, in a strained atmosphere of widespread Ministry scepticism of local planning, even the revered Abercrombie was not immune from criticism. One of the Planning Ministry's Assistants, Gatliff, wrote of the Abercrombie/Lutyens plan of Hull that 'generally, it seems to me a tragedy both for Hull, Sir Patrick Abercrombie and planning generally that he ever went near the place, and the sooner Hull gets away from his wilder ideas and faces up to the practical job of replanning . . . in a sound, decent, ordinary way the better' (letter by Gatliff, 14 Feb, 1946, NA HLG 79/226).

Much of the post-war built form of both Coventry and Birmingham continues to be criticised in public representations – including by some of the respondents contributing to this study. Politicians and popular media tend to interpret these reconstructed cities as being well intentioned, but ultimately 'failed', attempts to implement unified architectural concepts; the implementation of these ideas has been interpreted by some as having negatively influencing the cities' economic situation and that continues to detract from the overall quality and experience of the environment. For example, a 2013 piece in the *Economist* presented a very mordant view of Birmingham's post-war rebuilding programme and the city's poor economic performance, suggesting that 'Victorian New Street Station was knocked down and replaced with a grim, urine-soaked box; the Edwardian shopfronts on New Street were replaced with plastic and concrete. Over time that helped to turn Birmingham from the country's most successful big cities into one of its least' (*The Economist*, 28 May, 2013). This is a very sanitised view of the past and the realities of an ageing and bomb-damaged city, ignoring the macro-economic issues such as globalisation that have adversely affected the city's economy.

Gibson and Manzoni's pursuit of well-intentioned or even rationally ordered cities endorsed the evisceration of familiar landmarks to make way for ill-conceived shopping spaces and inhumane road schemes. These 'new' city centres were received, at least after a passage of time, by citizens with a sense of indifference or even enmity. For example, in 2011, Simon Jenkins, as the Chairman of the National Trust, re-affirmed this popular narrative suggesting that the post-war 'comprehensive clearance of much of central Birmingham, Manchester and Liverpool removed familiar landmarks and replaced them with a blighted townscape of concrete and tarmac' (Jenkins, 2011, page 262). There is some considerable veracity to these claims – and it would certainly be very difficult and unwise to argue stridently against such criticisms, even when faced with the temptation to 'blame' the physical failings of individual post-war buildings/infrastructure projects, rather than the 'planners' broader ideas behind their production, for certain design flaws and social problems.

Recent interpretations of Gibson and Manzoni's influence highlight their role in pursuing ambitious modernist replanning ideas conceived from an 'elevated' and privileged (or even 'lofty') position. It is difficult to present a compelling countervailing case against such a general assessment: certainly, their professional backgrounds lend credence to this argument with both men being middle-class, educated, urbane and in Manzoni's case, 'well-connected' in Ministerial (both Planning and Transport) circles. Ideas relating to the replanning of central Coventry, particularly in relation to the displays used during the city's large-scale public exhibitions, and the diagrams used to show plans for Birmingham's five 'New Towns', were informed by a detailed analysis of assiduously collected spatial data related to land-use patterns and environmental quality. Even though Manzoni refused to acknowledge the direct influence of the Bournville Village Trust publication, *When we build again* (1941), on Birmingham's approach to post-war replanning, it did include a detailed social mapping work of Birmingham and its wider hinterland. Manzoni also was involved in the later West Midlands Group publication, *Conurbation* (1948), which, though regional in scope, used statistical data, thematic maps and photographs. In short, like other cities, these representations sought to encourage a belief in the evidence-based 'science' of planning; that solutions can come 'from above', where imagined future possibilities involved moving away from the 'unordered' city of the late-nineteenth and early-twentieth century. Views 'from below' were marginalised in both cities: as elsewhere in this period, there is very little evidence that public opinion expressed at any of the numerous plan exhibitions actually changed the plans.

Both cities set aside areas of the broad zoning within the ring roads for shopping, civic and industrial use. In Coventry, exceptionally among British cities wrestling with the issue of post-war city reconstruction, not only did the City Architect prepare the plan but the City Architect's department designed also by far most of the buildings (Gould and Gould, 2016). Coventry has retained a clear sense of Gibson's vision despite the recent changes made to the built form. Birmingham, however, did not enjoy such a coherent approach. Today, much of the post-war rebuilding is being rebuilt again, or being threatened with plans for redevelopment. Although some will survive, either through preservation (e.g. the nationally Listed Rotunda), others have suffered the indignity of being demolished or radically altered. Key elements of Coventry's post-war built forms have been retained; the ring road, the zones of commercial and civic space and, at the centre of city, the pedestrian precinct with its alignment focussed on the Cathedral – although these principles are also under clear risk from the forces of commercial development. The national Listing of some elements of this internationally known flagship rebuilding project has come surprisingly late.

Coventry's was the first and the most widely publicised of the reconstruction plans; and, as the first provincial city to be comprehensively damaged in the blitz of 1940, and with a City Architect and his team already in place and considering its reconfiguration, it was well placed to proceed with the replanning of the centre. In terms of its scale and its comprehensiveness, Coventry can be situated somewhere between Plymouth and Exeter in terms of 'boldness'. Though, as Gould and Gould (2016) note, neither Abercrombie nor Sharp openly acknowledge Coventry – or Gibson – as being an influence on their planning ideas. Gibson's recognised inspirations were drawn from the modern ideas of Le Corbusier, but also Lewis Mumford. However, the balanced axial *Beaux-Arts* approach supported by Abercrombie and the Liverpool School of Architecture is apparent both in the layout of the Upper and Lower Precinct and in the formal designs for the Civic Quarter, and there was certainly an appreciation of projects in other places that informed the development of Coventry's plan (Johnson-Marshall, 1966). Notable among these was the Lijnbaan in Rotterdam, which bears a strong resemblance to Coventry's Shopping Precinct, a pedestrian shopping street crossed by trafficked streets and the pedestrianisation of the main shopping street, the arcading of the shops and elevated walkways. Birmingham's ultra-modern 'Brave New World' envisaged by Paul Cadbury (1952) was never truly realised. Like Coventry, the city received ring road(s); new housing with modern facilities replaced slums; sizeable shopping centres and office blocks transformed the city centre, all without a reconstruction plan to work form. Manzoni was the overseer of this process, but even he could not proceed single-handedly.

It might be provocatively suggested that too much current debate focuses on how power is wielded by planners and other decision-makers and how such power is resisted 'on the ground' by those groups and individuals in a select group of cities. Yet there is abundance of scholarly work that demonstrates the role played by macro and micro forces in shaping development activity. Developers, finance houses, the church, merchant banks, trading houses, family firms, landowners, ethnic diasporas, politicians, estate agents, architects, contractors and citizens of different racial, sexual or class orientation, all influence 'what gets built, used and conserved'. In Coventry and Birmingham, both Gibson and Manzoni had experience of 'their' cities and the areas that needed replanning before the destruction of the Blitz, but they certainly did not act alone in reshaping their cities. Gibson both created and pursued the early realisation of the plan; though he left Coventry in 1955 and the subsequent City Architects, Arthur Ling and Terence Gregory, added their own individual interpretation to the original plan.

Coventry purposely chose not to select an eminent external planning consultant (unlike Plymouth or Exeter). Donald Gibson both received considerable support from his (newly elected Labour) city council and fellow officers and he assumed the powerful status of Architect and City Planning Officer. Furthermore, from his appointment in 1938, Gibson had had opportunity to consider how Coventry could be re-configured and, to this end, proceeded to assemble a team of enthusiastic architect-planners (cf. Johnson-Marshall, 1966; Tiratsoo, 1990), many of whom were Liverpool-trained (at the Department of Civic Design and under Professor Charles Reilly). Birmingham's approach to reconstruction was similarly a very much in-house and local affair. Like Gibson, Manzoni was also educated at Liverpool University, though as an engineer rather than an architect. Manzoni could influence decision-making having managed to influence the provisions of the 1944 Planning Act to the city's advantage. He could not, and did not, hold autonomous power in bringing about reconstruction: personalities such as the forthright Frank Price pushed for and achieved, construction, while Birmingham-based property developer, Jack Cotton, was also influential in driving office development in the city centre.

In Coventry, like Exeter and Plymouth, the City Architect enjoyed having equal standing with the other chief officers, and his ideas could be heard. For Birmingham, however, a post of city architect was created in 1952, but the first appointee, Sheppard Fidler, was largely subservient to Herbert Manzoni; and it was Fidler – again a graduate from Liverpool University with a 'First' in Architecture – who demonstrated any consideration towards historic buildings and aesthetic concerns. Personal tensions between Manzoni and Fidler also hampered efforts to co-ordinate present a coherent approach to Birmingham's reconstruction. Leslie Ginsberg (founding Head of the Birmingham School of Planning) suggested that the Council's planning functions were dispersed between too many disparate departments, with the redevelopment section having a very small pool of trained planners and no architect (Ginsberg, 1959). Even though Fidler was partially successful in bringing a more sensitive design approach in residential redevelopment, he also ran into the difficulty when the then Labour leader in the early 1960s, Henry Watton, pushed for increased numbers of dwellings. This affected the process of site development and contractual policy (see Glendinning and Muthesius, 1994, page 248).

A timely reassessment?

The narratives of those residents experiencing processes of post-war reconstruction of Coventry and Birmingham go some way to bring into question the relationship between the vision of modernist planner-architects or even the planning frameworks that lay behind it (Gold, 1997) and the resultant, and now much-derided, built form. These are new perspectives. In Coventry's case, several respondents argued that there were entirely understandable reasons why Gibson's plans were conceived, suggesting that the city had a high proportion of incomers, attracted by the city's history of manufacturing employment, and the materiality of the 'new' city centre needed to reflect these changing circumstances (see also Campbell, 2004). In Birmingham, Manzoni and colleagues worked hard to provide improved living conditions, better facilities for business and shopping and to cope with increases in motorised traffic. These planning interventions were based on the systematic and detailed collection and thorough analysis of data. In a broader context, there are similarities here with the work of French interwar social scientists, and they used aerial photography to assess the socio-spatial characteristics of places, which then helped them promote sociological perspectives and to gain an 'insight into the social trends that existed below the surface' (Haffner, 2013, page 138). Perhaps there is too much analysis of the stifling impacts of planning activity and that there is nothing intrinsically 'wrong' with setting out visions for more sanitary, cleaner, healthier, safer, self-sustaining, greener and culturally enriched urban spaces replete with ample employment opportunities (Hall, 2014).

Ideas regarding reconstruction were often shaped by personal, economic, political or professional tensions at the national and local level. They were the product of many minds and they altered over time; they were also filtered through consultation with local councillors, local officers, developers and the wider public. This points towards the ways in which political elites (both locally and nationally) and developers appropriated the vernacular of modernism to provide a response to political pressures for slum clearance and city-centre redevelopment (cf. Flinn, 2012, 2013; Greenhalgh, 2017). The fact remains, though, that plans do not always determine how the city is used; new 'lived' spaces often emerge as people adapt the city to their own ends. The narratives used in this study highlight how ongoing urban modernisation might be a contested process; people's practical negotiation of buildings, sites and infrastructure sometimes coalesce with certain planning ideas as well having the potential to escape, or even emerge from, the 'planned-for' city.

The value of recollections and memories

This research has considered the potential of an oral history-type approach to heed the recent calls within planning history for a more nuanced interpretation of how the inhabitants of cities reacted to the transformation of the (post-war) townscape from the ground. Situated within this context, therefore, this research draws on the embodied understandings of post-war change, embedded in the narratives of those people who have lived through and *consumed* the post-war reconstruction of Coventry and Birmingham. Rather than solely relying on an 'official' narrative assembled through council minutes, or through a thorough analysis of national and local archival material, the experience of reconstruction is being embedded in material environments. For example, it has been seen through interviewing 'in-the-field' that the tangible and intangible aspects of place – buildings, landmarks, but also smells, sounds and absent sites – can be used to stimulate memories, stories, discussions and recollections of sensory experiences as interviews moved around the two city centres.

Drawing attention to the contested nature of the modernisation using people's memories can contribute meaningfully to a broader historical reassessment of the modernist built environment. Discussions over the value of post-war heritage continue to expand, and it may be the case that urban decision-makers and the public become increasingly receptive to some of the design principles of post-war rebuilding, or to its societal value and significance. There is national and local recognition of the historical significance of the Gibson plan for Coventry. Yet the prevailing local narrative is that post-war redevelopment fractured the pre-war (prosperous) historical city, becoming, as While (2006, page 2414) suggests 'part of, rather than simply reflecting, the city's decline in the 1970s and 1980s'. Gibson's plan endures, and residents have an ongoing engagement with the city centre. A strong case could be made that Coventry City Council should push for a central precinct conservation area based around the preservation of Gibson's plan (cf. Gould and Gould, 2016).

Similarly, in Plymouth, the 'Abercrombie townscape' of grand boulevards and refined Portland stone facades might (but, in practice, does not) have a certain appeal in terms of conservation-led regeneration. Consideration is given towards the conservation of the central area of Coventry, especially the need to 'protect and adapt existing buildings, as much as possible' (Coventry City Council, 2012, page 28) in the city centre, especially Broadgate, the Upper and Lower Precincts, the south end of Smithford Way, some of Market Way including the covered market, Bull Yard and parts of Hertford Street and Corporation Street together with the Civic and University quarters. As Gould and Gould (2016) also point out, attention might focus on the reinstatement of Market Street as a 'trafficked' street, while the ring road could be returned to grade and the legibility and connectivity between the centre and the outlying areas improved/repaired. Like Bristol's Broadmead, Birmingham's post-war piecemeal redevelopment lacks the coherence of Plymouth's grand plan or the originality of Coventry's precincts and the absence of an overarching plan has, arguably, allowed for selected elements of the post-war built form to be more easily removed: or, put another way, there is far less persistence of Manzoni's planning ideas. Unless, of course, one argues that his planning ideas focussed on accepting change, impermanence and flexibility for urban spaces and services!

However, respondents generally recorded lukewarm or ambivalent reception to the broad principles of post-war conservation. There was support for the recent regeneration initiatives, including the removal of what some perceived to be the out-dated post-war townscape; and yet, as with other cities, respondents hinted that some of the *ideas* and values associated of the modernist environment are being lost in the contemporary market-led redevelopment of the city centre (Campkin, 2014). A case might also be made for rediscovering its original sense of optimism associated with some post-war architectural schemes – especially its expression of a belief in the possibility of social progress and human betterment. In other ways, however, such an approach transcends recent critiques of urban regeneration.

Unlike other studies that tend to position individuals' nostalgias as being 'antagonistic' towards the official attempts to reshape aspects of the (urban) past, we see this relationship as being more complex (see also Adams and Larkham, 2016). We find great difficulty in arguing against the idea that sweeping and rapid large-scale urban change disrupts the lives, histories, routines and general experience of everyday life for users of urban space. Disruption is inevitably painful; although it can (and is increasingly being used so in contemporary governance and business) be used as a creative process. However, paying attention to people's embodied engagement with the material environment also reveals how individuals' affectionate memories were also triggered by some respondents' accounts of what *was* lost.

Feelings of 'loss' provoked by the forces of modernity can hold powerful affective charges.

What is less reasonable, however, is to accept that the personalised narratives collected in this study uncritically and blindly argue for retrofitting a nostalgic or artificial version of the past – of what might have been – into future designs. Sensitivity to location and context is key, too. Some people would interpret the physical reconstruction of certain buildings or monuments may be associated with efforts to resuscitate 'darker' aspects of a city's history. Nevertheless, this work therefore raises questions in relation to debates about collective responsibility for and the maintenance of the post-war built heritage for *all* people and how to incorporate wider active citizenship in planning decisions. Though, as the European Framework Convention on the Value of Cultural Heritage for Society makes clear, all people have the responsibility to respect the cultural heritage of others as much as their own.

Where next?

This work re-evaluated existing evidence from two contrasting cities and sought experiences from small groups of long-term residents. From a methodological perspective, more respondents could have been sourced for the go-alongs, to provide a richer set of experiences from, for example, different socio-economic groups. Experiences of those who have moved in to these cities could be valuable, as they have other experiences from which to draw comparisons. Wherever practicable, future research could also employ similar methods to capture *younger* residents' feelings and attitudes towards elements of the post-war landscape: to date, this is a perspective that is underdeveloped and requires some further practical and theoretical consideration. One immediate advantage, however, given the 'active' nature of walking, is that these interviews might appeal to those more physically able and who could be willing to provide rich perspectives. The go-alongs can help provide a stimulus and motive for submerged recollections, triggered, sometimes unpredictably, by encounters of objects, buildings, specific smells, sights or sounds and absent sites. While conducting these guided walks in the 'present' holds rich potential for practitioners who seek to re-awaken and then look to incorporate, positive memories from people of different ages, genders, sexualities, religions, races and socio-economic statuses into official decision-making processes. Additionally, there is considerable potential for these types of locally generated narratives to be captured, digitised and then (re)presented in GIS in such a way that brings into question the power and knowledge inherent in the 'planners' use of maps to 'plan for' places (see Perkins and Dodge, 2012). Both sedentary and walking methods have great potential significance in terms of transferring ideas into practice. Public decision-making has a responsibility for be sympathetic to the preferences of local populations, linking the past to the future in a sensitive way. This study therefore points to how other researchers might unpick the ambivalent relationship between modernity, loss and the built form and how people's engagement with urban materiality can contribute to the potential adaptive psychological functioning, especially among the elderly.

Although much has been written on post-Second World War reconstruction, Diefendorf (2012) has argued that comparative studies can help inform decision-makers in cities that have in recent years also suffered wartime destruction, and this work can also provide valuable insights for cities seeking to rebuild after major natural disasters (Bold *et al.*, 2018). It develops a new perspective towards urban physical and social resilience. These studies are encouraging; therefore, the comparative analysis of Coventry and Birmingham – and the blending of archival research *and* oral history-type perspectives from residents – could contribute to the wider study of other European cities' approaches to post-war reconstruction. Distilling the key messages from other case studies allows researchers and decision-makers to identify issues and principles on the way to developing future guidelines offering pragmatic ways forward, to help shape a positive response to future change, whether itself catastrophic or induced by catastrophes.

Note

1 As Jencks (1977, page 9) famously pointed out in his comment on the precise time of the demolition of the 'failed' Pruitt Igoe municipal housing estate: 'modern architecture died in St Louis, Missouri, on July 15, 1972, at 9:32 pm'. He was rhetorically using this example to date the origin of 'post-modernism'.

References

No full reference is given here to short news items of letters to the editor in newspapers or professional journals: these can be found from the edition date and details given in the text.

Abercrombie, P. (1945) *The Greater London Plan 1944* HMSO, London.

Abercrombie, P. (1949) 'Planning in Britain', *Architecture* 37, (1), pages 10–13.

Abercrombie, P. and Matthew, R.H. (1949) *The Clyde Valley Regional Plan 1946* HMSO, Edinburgh.

Adams, D. (2011) 'Everyday experiences of the modern city: Remembering the post-war reconstruction of the city', *Planning Perspectives* 26, (2), pages 237–360.

Adams, D. and Larkham, P.J. (2013) 'Bold planning, mixed experiences: The diverse fortunes of post-war Birmingham', in Clapson, M. and Larkham, P.J. (eds.) *The Blitz and Its Legacy: From Destruction to Reconstruction* Ashgate, Aldershot.

Adams, D. and Larkham, P.J. (2016) 'Walking with the ghosts of the past: Unearthing the value of residents' urban nostalgias', *Urban Studies* 53, (10), pages 2004–2022.

Addison, P. (1975) *The Road to 1945: British Politics and the Second World War* Pimlico, London.

Aldous, T. (1975) *Goodbye, Britain?* Sidgwick and Jackson, London.

Amery, C. and Cruickshank, D. (1975) *The Rape of Britain* HarperCollins, London.

Architect and Building News (1941) 'New Plan for Coventry', 21 Mar, page 192.

Architect and Building News (1943) 8 Oct, page 22.

Architect and Building News (1955) 'When is a Bull Ring a white elephant', 8 Dec, page 1074.

Architect and Building News (1959) 'The City of Birmingham rebuilds', 15 Apr, pages 470–482.

Architects' Journal (1941) 'Plan for the New Coventry', 24 Apr, pages 278–281.

Architect's Journal (1953) 'Coventry', 107, 8 Oct.

Architects' Journal (1975) 'First stage of the Spon Street scheme, Coventry', 26 Mar, page 655.

Architectural Design (1958) 'Coventry rebuilds', 28, pages 473–474.

Architectural Review (1941) 'CCA scrapbook of architect's department news cuttings', Oct, page 110.

Architectural Review (1943) 'Rebuilding Britain', 93, (556), Apr.

Ashworth, G. and Tunbridge, J. (1996) *Dissonant Heritage: The Management of the Past as a Resource in Conflict* Wiley, London.

Ashworth, W. (1954) *Genesis of Modern British Town Planning* Routledge and Kegan Paul PLC, London.

Atkinson, H. (2012) *The Festival of Britain: A Land and Its People* I.B. Tauris, London.

Baldwin, R. (2005) *Visions of Reconstruction 1940–1948* Thomas Telford, London.

Barnett, C. (1986) *The Audit of War: The Illusion and Reality of Britain as a Great Nation* Macmillan, London.

Bartram, R. and Shobrook, S. (2001) 'Body beautiful: Medical aesthetics and the reconstruction of urban Britain in the 1940s', *Landscape Research* 26, pages 119–135.

Beanland, C. (2013) 'The man who built Brum: A lament for the demise of John Madin's Brutalist Birmingham', *The Independent* 12 Sept.

Beaven, J. and Griffiths, J. (1999) 'The Blitz, civilian morale and the city: Mass-observation and working-class culture in Britain, 1940–1941', *Urban History* 26, pages 71–88.

Berman, M. (1983) *All That Is Solid Melts into Air: The Experience of Modernity* Verso, New York.

Beveridge, W. (1942) *Social Insurance and Allied Services* HMSO, London.

Bianconi, M. and Tewdwr-Jones, M. (2013) 'The form and organisation of urban areas: Colin Buchanan and traffic in towns 50 years on', *Town Planning Review* 84, (3), pages 313–336.

Birmingham City Council (1956–1957) *City of Birmingham Inner Ring Road* Public Works Committee, City Council, Birmingham.

Birmingham City Council (1957) *The Redevelopment of the Central Areas* unpublished notes for professional visitors, City Council, Birmingham.

Birmingham City Council (1989a) *Developing Birmingham 1889–1989* City Council, Birmingham.

Birmingham City Council (1989b) *The Highbury Initiative City Challenge Symposium* City Council, Birmingham.

Birmingham City Council (1995) *Architecture and Austerity: Birmingham 1940–1950* Department of Planning and Architecture, City Council, Birmingham.

Birmingham City Council (1999) *Regeneration through Conservation: Strategy* Birmingham.

Birmingham City Council (2011) *Big City Plan* City Council, Birmingham.

Birmingham Evening Mail (1959) 'The new Birmingham'.

Birmingham Evening Mail (1965) 'City Critics "live in fairy land"', 8 Apr.

Birmingham Evening Mail (1969) 'Letter', 30 Oct.

Birmingham Evening Mail (2003) 'It's a ghost city', 30 Aug.

Birmingham Mail (2012) 'First silver panels installed in New Street revamp', 15 Dec.

Birmingham Post (1941) 'Letter by W. Randolph', 28 Jan.

Birmingham Post (2012a) 'Birmingham Central Library demolition gets go-ahead', 2 Dec.

Birmingham Post (2012b) 'Developments key in helping Coventry to realise its potential', 27 Feb.

Black, H. (1957) *History of the Corporation of Birmingham* (Vol. 6) City Council, Birmingham.

Bold, J., Larkham, P.J. and Pickard, R. (2018) *Authentic Reconstruction* Bloomsbury, London.

Borg, N. (1973) 'Birmingham', in Holiday, J. (ed.) *City-Centre Redevelopment* Charles Knight, London.

Bournville Village Trust (1941) *When We Build Again* BVT, Birmingham.

Briggs, A. (1952) *History of Birmingham, Vol. 2: Borough and City 1865–1938* Oxford University Press, London.

Brooke, S. (1992) *Labour's War: The Labour Party and the Second World War* Clarendon Press, London, pages 171–172.

Bruce, R. (1945) *First Planning Report to the Highways and Planning Committee of the Corporation of the City of Glasgow* Abstracted/reviewed in *Architects' Journal*, pages 63–66.

Buchanan, C. (1963) *Traffic in Towns* Penguin, Harmondsworth.

The Builder (1945) 'The ring road challenged', report of paper by L.B. Escrit to the Association for Planning and Regional Reconstruction 28 Dec, page 519.

Bullock, N. (2002) *Building the Post-War World: Modern Architecture and Reconstruction in Britain* Routledge, London.

Cadbury, P. (1952) *Birmingham: Fifty Years On* Bournville Village Trust, Birmingham.

Calder, A. (1991) *The Myth of the Blitz* Random House, London.

Calder, B. (2016) *Raw Concrete: The Beauty of Brutalism* Penguin Random House, London.

Campbell, L. (2004) 'The Phoenix and the city: War, peace and architecture', in Mac-Cormac, E. and Lovell, V. (eds.) *Phoenix: Architecture/Art/Regeneration* Black Dog, London.

Campbell, L. (2006) 'Paper dream city/modern monument: Donald Gibson and Coventry', in Whyte, I.B. (ed.) *The Man-Made Future: Planning, Education and Design in the Mid-Twentieth Century* Routledge, London.

Campkin, B. (2014) *Remaking London: Decline and Regeneration in Urban Culture* I.B. Taurus, London.

Carmona, M., Heath, T., Oc, T. and Tiesdell, S. (2009) *Public Places, Urban Spaces: The Dimensions of Urban Design* Architectural Press, London.

Carran, E. (1941) 'Letter to the editor', *The Builder* 160, (5113), 31 Jan, page 124.

Castells, M. (2000) *The Rise of the Network Society* (2nd edition) Blackwell, Oxford.

Caulcott, T. (2004) 'Manzoni, Sir Herbert John Baptista', in Matthew, H.C.G. and Harrison, B. (eds.) *Oxford Dictionary of National Biography* 36 Oxford University Press, Oxford.

Chan, W.F. (2005) 'Planning at the limit: Immigration and post-war Birmingham', *The Journal of Historical Geography* 31, (3), pages 513–527.

Chapman, D. (2005) 'Knowing and unknowing: Development and reconstruction planning in Malta from 1943', *Journal of Urban Design* 10, (2), pages 229–252.

Cherry, G. (1981) *Pioneers in British Planning* Architectural Press, London.

Cherry, G. (ed.) (1988) *Cities and Plans: The Shaping of Urban Britain in the Nineteenth and Twentieth Centuries* Edward Arnold, London.

Cherry, G. (1989) 'Lessons from the past: Abercrombie's Plymouth', *Planning History* 11, (3), pages 3–7.

Cherry, G. (1994) *Birmingham: A Study in Geography, History and Planning* Wiley, Chichester.

Cherry, G. and Penny, L. (1986) *Holford: A Study in Architecture, Planning and Civic Design* Mansell Publishing, London.

Chinn, C. (1999) *Homes for People: Council Housing and Urban Renewal in Birmingham, 1849–1999* Brewin Books, Studley.

Chinn, C. (2013) 'Change came fast with Herbert Manzoni', *Birmingham Mail* 18 May.

Church of England (2011) 'May God bless this great city', *News from the Diocese of Coventry* (available at: www.coventry.anglican.org).

Clapson, M. and Larkham, P. (eds.) (2013) *The Blitz and Its Legacy: From Destruction to Reconstruction* Ashgate, Aldershot.

Clawley, A. (2011) *John Madin* RIBA Publishing, London.

Clayton, N. and Sivaev, D. (2013) 'Driving growth: Supporting Business Innovation in Coventry and Warwickshire, report for the Centre for Cities (available at: www.centreforcities.org/assets/files/2013/13-05-17-Coventry-and-Warwickshire.pdf).

Clitheroe, G. (1942) *Coventry under Fire, November 1940–April 1941* The British Publishing Company Ltd, London.

Colby, J. (1964) *Tear Down to Build Up* Cadmus, Eau Claire.

Cook, I., Ward, S. and Ward, K. (2013) 'A Springtime journey to the Soviet Union: Post-war planning and policy mobilities through the Iron Curtain', *International Journal of Urban and Regional Research* 34, (2), pages 415–420.

Corbett, N. (2004) 'Renaissance in Birmingham', in Corbett, N. (ed.) *Transforming Cities: Revival in the Square* RIBA, London.

Coulson, A. (2003) 'Review of Price', F. (2002) *Being There* Upfront Publishing, Kibworth, in *Local Government Studies* 29, (1), pages 127–129.

Coventry City Council (1945) *The Future Coventry* City Council, Coventry.

Coventry City Council (1958) *Development and Redevelopment in Coventry* Public Relations Department, Coventry.

Coventry City Council (2012) *Proposed Core Strategy* Coventry City Council, Coventry.

Coventry Evening Telegraph (1941) 'Letter', 7 Jan, page 6.

Coventry Evening Telegraph (1948) 'The elephant mast in the Broadgate', 22 Apr, page unknown.

Coventry Evening Telegraph (2012) 'Historic city centre rebuild remains ambitious in economic climate', 28 Feb.

Coventry Evening Telegraph (2013a) 'Coventry council offices moving to Friargate development as part of £59m masterplan for city centre', 21 May.

Coventry Herald (1940a) 'Letter', 4 May, page unknown.

Coventry Standard (1940b) 'Letter', 18 May, page unknown.

Cowan, S. (2013) 'The people's peace: The myth of wartime unity and public consent for town planning', in Clapson, M. and Larkham, P.J. (eds.) *The Blitz and Its Legacy: From Destruction to Reconstruction* Ashgate, Aldershot.

Cowles, B., Barron, J., Bishop, F., Piggott, L., Watson, A., Alford, A., Dallaway, J. and Warren, D. (1975) 'Discussion: Birmingham inner ring road', *Proceedings of the Institute of Civil Engineers* 58, pages 453–456.

Cullingworth, J.B. (1975) *Reconstruction and Land Use Planning 1939–1947* HMSO, London.

Curl, J.S. (1968) a series of articles published in *The Oxford Mail* 25 Oct, entitled 'The Erosion of Oxford'.

Daily Mail (1941) 'City of open spaces: New Coventry takes shape', 24 Feb, page unknown.

Davies, J.G. (1972) *The Evangelistic Bureaucrat* Taylor & Francis, London.

Degen, M. (2017) 'Urban Regeneration and "Resistance of Place": Foregrounding Time and Experience', *Space and Culture* 20, (2), pages 141–155.

Degen, M. and Rose, G. (2012) 'The sensory experiencing of urban design: The role of walking and perceptual memory', *Urban Studies* 49, (15), pages 3271–3287.

Demidowicz, G. (2002) 'Coventry's Phoenix initiative', *Context* 76, Sept.

Dennis, N. (1972) *Public Participation and Planner's Blight* Faber and Faber, London.

Department for Communities and Local Government (2012) *National Planning Policy Framework* DCLG, London.

DeSilvey, C. and Edensor, T. (2013) 'Reckoning with ruins', *Progress in Human Geography* 37, pages 465–485.

Diefendorf, J.M. (1989) 'Urban reconstruction in Europe after World War II', *Urban Studies* 26, (1), pages 128–143.

Diefendorf, J.M. (ed.) (1990) *Rebuilding Europe's Bombed Cities* Macmillan, Basingstoke.

Diefendorf, J.M. (2012) 'Rebuilding the cities: Destroyed in World War II: Growing possibilities for comparative analysis', paper presented at the Conference of the European Association for Urban History, Prague, on 'Cities and Societies in Comparative Perspective'.

Dix, G. (1981) 'Patrick Abercrombie', in Cherry, G. (ed.) *Pioneers in British Town Planning* Architectural Press, London.

Douglas, A. (1983) *Coventry at War* Brewin Books, Studley.

Drive (1971) 'With this ringway . . .', Autumn, pages 50–53.

Drozdzewski, D., De Nardi, S. and Waterton, E. (eds.) (2016) *Memory, Place and Identity: Commemoration and Remembrance of War and Conflict* Routledge Press, London.

DTZ (2013) *City Centre Retail Assessment 2013* Prepared for Birmingham City Council, Birmingham.

Durkheim, E. (1893 [1997]) *The Division of Labor in Society* New York, Free Press.

Düwel, J. and Gutschow, N. (2013) *"A Blessing in Disguise" War and Town Planning in Europe 1940–1945* DOM, Berlin.

Dyckhoff, T. (2006) 'Let's move to . . . Coventry', *The Guardian* 11 Mar.

Dyos, H. (ed.) (1968) *The Study of Urban History* Edward Arnold, London.

The Economist (2013) 'How to kill a city', 28 May.

Elkes, N. (2018) '"We will demolish every council high rise tower block in Birmingham", say Tories', *Birmingham Mail* 27 Mar.

Emery, J. (2006) 'Bullring: A case study of retail-led urban renewal and its contribution to city centre regeneration', *Journal of Retail and Leisure Property* 5, pages 121–133.

Esher, L. (1981) *A Broken Wave: The Rebuilding of England 1940–1980* Allen Lane, London.

Essex, S. and Brayshay, M. (2005) 'Town versus country in the 1940s', *Town Planning Review* 76, pages 239–263.

Essex, S. and Brayshay, M. (2007) 'Vision, vested interest and pragmatism: Who re-made Britain's blitzed cities', *Planning Perspectives* 22, (4), pages 417–441.

Essex, S. and Brayshay, M. (2008) 'Boldness diminished? The post-war battle to replan a bomb-damaged provincial city', *Urban History* 35, (3), pages 437–461.

Essex, S. and Brayshay, M. (2013) 'Planning the reconstruction of war-damaged Plymouth, 1941–1961: Devising and defending a modernisation agenda', in Clapson, M. and Larkham, P.J. (eds.) *The Blitz and Its Legacy: From Destruction to Reconstruction* Ashgate, Aldershot.

Fischer, K.F. and Larkham, P.J. (2018) 'Coventry: A model of modernist reconstruction', in Fischer, K.F. and Altrock, U. (eds.) *Windows Upon Planning History* Routledge, Abingdon, Oxon.

Flinn, C. (2012) '"The city of our dreams"? The political and economic realities of rebuilding Britain's blitzed cities, 1945–1954', *Twentieth Century British History* 23, (2), pages 221–245.

Flinn, C. (2013) 'Reconstruction constraints: Political and economic realities', in Clapson, M. and Larkham, P.J. (eds.) *The Blitz and Its Legacy: From Destruction to Reconstruction* Ashgate, Aldershot.

Flynn, N. and Taylor, A. (1986) 'Inside the rust-belt: An analysis of the decline of the West Midlands economy', *Environment and Planning A* 18, pages 865–900.

Forshaw, J.H. and Abercrombie, P. (1943) *County of London Plan* Macmillan, London.

Foster, A. (2005) *Birmingham* Pevsner Architectural Guides Yale University Press, New Haven.

Foster, A. (2009) 'Birmingham Central Library's final chapter', *BD Magazine: Public Sector* July (available at: www.m.bdonline.co.uk).

Forster, E.M. (1937) 'Havoc', in Williams-Ellis, C. (ed.) *Britain and the Beast* J.M. Dent and Sons, London.

Fox, L. (1947) *Coventry's Heritage: An Introduction to the History of the City* Coventry Evening Telegraph, Coventry.

Fry, M. (1941) 'The new Britain must be planned', *The Picture Post* 4 Jan, pages 16–19.

Fyfe, N. (1996) 'Contested visions of a modern city: Planning and poetry in postwar Glasgow', *Environment and Planning A* 28, (3), pages 345–367.

Galloway, T.G. and Mahayni, R.G. (1977) 'Planning theory in retrospect: The process of paradigm change', *Journal of the American Institute of Planners* 43, (1), pages 67–71.

Gardiner, J. (2010) *The Blitz* Harper Press, London.

Gibson, D. (1940a) 'Planning post-war reconstruction', *Municipal Journal and Local Government Administrator* 13 Dec, pages 1583–1584.

Gibson, D. (1940b) 'Post-war civil development', *Camera Principis* (53), pages 2–3.

Gibson, D. (1941a) 'Preface', in Phillips, H.S. (ed.) *Town Development* self-published, Guildford, Surrey page 3.

Gibson, D. (1941b) 'Some matters concerning post-war reconstruction', address to the AA Annual General Meeting, 24 Jan, *Architectural Association Journal* Feb, pages 69–76.

Gibson, D. (1947) 'Architects to public authorities', *Journal of the Royal Institute of British Architects* June.

Gill, R. (2004) 'From the Black Prince to the Silver Prince: Relocating mediaeval Coventry', *Twentieth Century Architecture* 7, pages 61–86.

Gillon, J. and McDowell, D. (2012) *Edinburgh's Post-War Listed Buildings* City of Edinburgh Council, Edinburgh.

Ginsberg, L. (1959) 'Town planning or road building?', *Architects' Journal* 130, (3), Oct, pages 288–294.

Glendinning, M. and Muthesius, S. (1994) *Tower Block: Modern Public Housing in England, Scotland, Wales and Northern Ireland* Yale University Press, New Haven.

Goebel, S. (2011) 'Commemorative cosmopolis: Transnational networks of remembrance', in Goebel, S. and Keene, D. (eds.) *Post-War Coventry Cities into Battlefields: Metropolital Scenarios, Experiences and Commemoration of Total War* Ashgate, Aldershot.

Gold, J.R. (1997) *The Experience of Modernism: Modern Architects and the Future City, 1928–1953* Spon, London.

Gold, J.R. (2007) *The Practice of Modernism: Modern Architects and Urban Transformation, 1954–1972* Routledge, London.

Gold, J.R. and Ward, S. (1994) 'We're going to do it right this time: Cinematic representations of urban planning and the British New Towns, 1939 to 1951', in Aitken, S. and Zonn, L. (eds.) *Place, Power, Situation and Spectacle: A Geography of Film* Rowman and Littlefield, Lanham, MA.

Gould, J. (2000) *Plymouth Planned: The Architecture of the Plan for Plymouth, 1943–1962* unpublished report commissioned by Plymouth City Council.

Gould, J. and Gould, C. (2016) *Coventry: The Making of a Modern City 1939–1973* Historic England, Swindon.

Greenhalgh, J. (2017) *Reconstructing Modernity: Space, Power and Governance in Mid-Twentieth Century British Cities* Manchester University Press, Manchester.

Gregory, T. (1973) 'Coventry', in Holiday, J. (ed.) *City-Centre Redevelopment* Charles Knight, London.

Gunn, S. (2011) 'The Buchanan Report, environment and the problem of traffic in 1960s Britain', *Twentieth Century British History* 22, (4), pages 521–542.

Gunn, S. (2018) 'Ring road: Birmingham and the collapse of the motor city ideal in 1970s Britain', *The Historical Journal* 61, (1), pages 227–248.

Gutschow, N. (2013) 'The blitz in England and the city of tomorrow', in Düwel, J. and Gutschow, N. (eds.) *War and Town Planning in Europe 1940–1945* DOM, Berlin.

Haffner, J. (2013) *The View from Above: The Social Science of Social Space* MIT Press, Cambridge, MA.

Hall, P. (1995) 'Bringing Abercrombie back from the shades: A look forward and back', *Town Planning Review* 66, pages 227–243.

Hall, P. (2014) *Cities of Tomorrow: An Intellectual History of Urban Planning and Design in the Twentieth Century* Wiley-Blackwell, Malden.

Hall, P. and Tewdwr-Jones, M. (2010) *Urban and Regional Planning* Routledge, London.

Hall, T. and Hubbard, P. (2014) '"Birmingham needs you/you need Birmingham": Cities as actors, actors in cities', in Conzen, M. and Larkham, P. (eds.) *Shapers of Urban Form* Routledge, Abingdon.

Hammerson plc. (2013) *Hammerson Acquires Additional Bullring Stake* (available at: www.hammerson.com/investors/rns/hammerson-acquires-additional-bullring-stake/).

Harris, R. and Larkham, P.J. (1999) *Changing Suburbs: Foundation, Form and Function* Routledge, New York.

Harvey, D. (1982) *The Limits to Capital* Blackwell, Oxford.

Harvey, D. (1989) *The Condition of Postmodernity* Blackwell, Oxford.

Harwood, E. (2003) *England: A Guide to Post-War Listed Buildings* Batsford, London.

Harwood, E. (2008) 'Neurath, Riley and Bilson, Pasmore and Peterlee', *Twentieth Century British Architecture 9*, pages 83–96.

Harwood, E. (2015) *Space, Hope, and Brutalism: English Architecture, 1945–1975* Yale University Press, London.

Hasegawa, J. (1992) *Replanning the Blitzed City Centre* Open University Press, Buckingham.

Hasegawa, J. (1999) 'Governments, consultants and expert bodies in the physical reconstruction of the City of London in the 1940s', *Planning Perspectives 14*, pages 121–144.

Hasegawa, J. (2013) 'The attitudes of the Ministry of Town and Country Planning towards blitzed cities in 1940s Britain', *Planning Perspectives 28*, (2), pages 271–289.

Haughton, T. and Martin, N. (2014) 'The long shadows and mixed modes of history: Concluding reflections on the aftermath and legacies of war', in Haughton, T., Martin, N. and Purseigle, P. (eds.) *Aftermath Legacies and Memories of War in Europe, 1918–1945–1989* Ashgate, Surrey, UK.

Healey, M.J. and Dunham, P.J. (1994) 'Changing competitive advantage in a local economy: The case of Coventry, 1970–1990', *Urban Studies 31*, (8), pages 1279–1302.

Hebbert, M. (1983) 'The daring experiment: Social scientists and land use planning in 1940s', Britain' *Environment and Planning B: Planning and Design 10*, (1), pages 3–17.

Hebbert, M. and Sonne, W. (2006) 'History builds the town: On the uses of history in twentieth-century city planning', in Monclus, J. and Gaurdia, M. (eds.) *Culture, Urbanism and Planning* Ashgate, Aldershot.

Hennock, E.P. (1973) *Fit and Proper Persons: Ideal and Reality in Nineteenth-Century Urban Government* Edward Arnold, London.

Henry, N. and Passmore, A. (2000) 'Rethinking global city centres: The example of Birmingham', *Soundings 13*, pages 60–66.

Hewison, R. (1995) *Culture and Consensus: England, Art and Politics Since 1940* Methuen, York.

Hewitt, K. (1983) 'Place annihilation: Area bombing and the fate of urban places', *Annals of the Association of American Geographers 73*, pages 257–284.

Hewitt, K. (1994) 'When the great planes came and made ashes out of our great city: Towards an oral geography of the disasters of war', *Antipode 36*, (1), pages 1–34.

Higgott, A. (1991–1992) 'A modest proposal: Abercrombie's County of London Plan 1943', *Issues in Architecture and Design 2*, pages 38–57.

Higgott, A. (2000) 'Birmingham: Building the modern city', in Deckker, T. (ed.) *The Modern City Revisited* Taylor & Francis, London.

Hodgkinson, G. (1970) *Sent to Coventry* Maxwell, London.

Holford, W. (1958) 'Coventry: Urban and suburban', *Architectural Design 28*, pages 480–481.

Holliday, J.C. (1973) *City Centre Redevelopment: A Study of British City Centre Planning and Case Studies of Five English City Centres* Charles Knight, London.

Hollow, M. (2012) 'Utopian urges: Visions for reconstruction in Britain, 1940–1950', *Planning Perspectives 27*, pages 569–585.

Holsten, J. (1989) *The Modernist City: An* Anthropological *Critique of Brasilia* University of Chicago Press, Chicago.

Holyoak, J. (2004) 'Street, subway and mall', in Kennedy, L. (ed.) *Remaking Birmingham: The Visual Culture of Birmingham* Routledge, Oxford.

Hopkins, E. (1989 [1998]) *Birmingham: The Making of the Second City, 1850–1939* Tempus, Stroud.

Hornsey, R. (2008) '"Everything is made of atoms": The reprogramming of space and time in post-war London', *Journal of Historical Geography* 34, (1), pages 94–117.

Hornsey, R. (2010) '"He who thinks, in modern traffic, is lost": Automation and the pedestrian rhythms of interwar London', in Edensor, T. (ed.) *Geographies of Rhythm* Ashgate, Aldershot, pages 99–112.

Howard, E. (1898) *To-Morrow: A Peaceful Path to Real Reform* Swan Sonnenschien, London.

Hubbard, P. (1996) 'Re-imagining the city: The transformation of Birmingham's urban landscape', *Geography* 81, pages 26–36.

Hubbard, P., Faire, L. and Lilley, K. (2003a) 'Contesting the modern city: Reconstruction and everyday life in post-war Coventry', *Planning Perspectives* 1, (8), pages 377–397.

Hubbard, P., Faire, L. and Lilley, K. (2003b) 'Memorials to modernity? Public art in the "city of the future"', *Landscape Research* 28, (2), pages 147–169.

Hubbard, P., Lilley, K. and Faire, L. (2004) 'Pacemaking the modern city: The urban politics of speed and slowness', *Environment and Planning D: Society and Space* 22, pages 273–294.

Jacobs, J.M. and Merriman, P. (2011) 'Practising architectures', *Social and Cultural Geography* 12, pages 211–222.

Jencks, C. (1977) *The Language of Post-Modern Architecture* Rizzoli, New York.

Jenkins, S. (2011) *A Short History of England* Profile Books Ltd., London.

Jewkes, J. (1948) *Ordeal by Planning* University of Michigan, Michigan.

Johnson-Marshall, P. (1958) 'Coventry: Test case for planning', *Official Architecture and Planning* May, pages 225–226.

Johnson-Marshall, P. (1966) *Rebuilding Cities* Edinburgh University Press, Edinburgh.

Jones, A. (2011) 'Urban impressions: Amazing Coventry' (available at: www.jones-theplanner.co.uk/2011/02/urban-impressions-amazing-coventry.html).

Jones, P. (1998) ' . . . a fairer and nobler City-Lutyens and Abercrombie's plan for the city of Hull 1945', *Planning Perspectives* 13, (3), pages 301–316.

Jones, P. (2004) 'Historical continuity and post-1945 urban redevelopment: The example of Lee Bank, Birmingham, UK', *Planning Perspectives* 19, pages 365–389.

Jones, P. (2008) 'Different but the same? Post-war slum clearance and contemporary regeneration in Birmingham, UK', *CITY* 12, (3), pages 356–371.

Judt, T. (2013) *Thinking the Twentieth Century* Random House, London.

Korn, A. (1953) *History Builds the Town* Percy Lund Humphries, London.

Kusenbach, M. (2003) 'Street phenomenology: The go-along as ethnographic research tool', *Ethnography* 4, pages 455–485.

Kynaston, D. (2007) *Austerity Britain, 1945–1951* Bloomsbury, London.

Kynaston, D. (2015) *Modernity Britain: 1957–1962* Bloomsbury, London.

Lagae, J. (2010) 'From "patrimoine partagé" to "whose Heritage"? Critical reflections on colonial built heritage in the city of Lubumbashi, Democratic Republic of the Congo', in Yacobi, H. and Fenster, T. (eds.) *Remembering, Forgetting and City Builders* Ashgate, London, pages 175–193.

Laing Development Company (1964) *The Bull Ring Centre, Birmingham* Promotional brochure Laing Development Company, London.

Lambert, R. (2000) 'Patrick Abercrombie and planning in Bath', *Bath History* 8, pages 172–196.

Larkham, P.J. (1997) 'Remaking cities: Images, control, and postwar replanning in the United Kingdom', *Environment and Planning B: Planning and Design* 24, (5), pages 741–759.

Larkham, P.J. (2002) 'Rebuilding the industrial town: Wartime Wolverhampton', *Urban History* 29, (3), pages 388–409.

Larkham, P.J. (2003) 'The place of urban conservation in the UK reconstruction plans of 1942–1952', *Planning Perspectives* 18, (3), pages 295–324.

Larkham, P.J. (2004) 'Rise of the 'civic centre' in English urban form and design', *Urban Design International* 9, (1), pages 3–15.

Larkham, P.J. (2006) 'The study of urban form in Great Britain', *Urban Morphology* 10 pages 117–141.

Larkham, P.J. (2007) *Replanning Birmingham: Process and Product in Post-War Reconstruction*, Faculty Working Paper series no. 2, School of Property, Construction and Planning, Birmingham.

Larkham, P.J. (2009) 'Thomas Sharp and the post war replanning of Chichester: Conflict, confusion and delay', *Planning Perspectives* 24, (1), pages 51–75.

Larkham, P.J. (2011) *Questioning Planning History* Faculty Working Paper Birmingham School of the Built Environment, Birmingham.

Larkham, P.J. (2013) 'Visions for a new Birmingham', in Larkham, P.J. (ed.) *When We Build Again* Routledge, Abingdon.

Larkham, P.J. (2014a) 'Municipal agency and post-catastrophe reconstruction', in Conzen, M. and Larkham, P. (eds.) *Shapers of Urban Form* Routledge, Abingdon.

Larkham, P.J. (2014b) 'Continual change: A century of urban conservation in England', *Centre for Environment and Society Research*, Working Paper series no. 21, Birmingham City University.

Larkham, P.J. (2016) 'Replanning post-war Birmingham process, product and longevity', *Architectura-Zeitschrift fur Geschichte der Baukunst* 46, (1), pages 3–26.

Larkham, P.J. and Adams, D. (2016) 'The un-necessary monument? The origins, impact and potential conservation of Birmingham Central Library', *Transactions of the Ancient Monuments Society* 60, pages 94–127.

Larkham, P.J. and Lilley, K.D. (2001) *Planning the City of Tomorrow: British Reconstruction Planning, 1939–1952 an Annotated Bibliography* Inch's Books, Pickering.

Larkham, P.J. and Lilley, K.D. (2003) 'Plans, planners and city images: Place promotion and civic boosterism in British reconstruction planning', *Urban History* 30, (2), pages 183–205.

Larkham, P.J. and Lilley, K.D. (2012) 'Exhibiting the city: Planning ideas and public involvement in wartime and early post-war Britain', *Town Planning Review* 83, (6).

Larkham, P.J. and Pendlebury, J. (2008) 'Reconstruction planning and the small town in early post-war Britain', *Planning Perspectives* 23, (3), pages 291–321.

Larkin, P. (1954) 'I remember, I remember', in Larkin, P. *The Less Deceived* Faber and Faber, London.

Le Corbusier (1929) *The City of Tomorrow and Its Planning* Architectural Press, London.

Lefebvre, H. (1991) *The Production of Space* Blackwell, Oxford.

Lewis, A. (2013) 'Planning through conflict: Competing approaches in the preparation of Sheffield's post-war reconstruction plan', *Planning Perspectives* 28, (1), pages 27–49.

Lilley, K. (2004) 'Experiencing the plan', in Larkham, P.J. and Nasr, J. (eds.) *The Rebuilding of British Cities: Exploring the Post-Second World War Reconstruction* School of Planning and Housing Working Paper series no. 90, University of Central England, Birmingham.

Llewellyn, M. (2003) 'Polyvocalism and the public: "doing" a critical historical geography of architecture', *Area* 35, pages 264–270.

Llewellyn, M. (2004) '"Urban village" or "white house": Envisioned spaces, experienced places, and everyday life at Kensal House, London in the 1930s', *Environment and Planning D: Society and Space* 22, pages 229–249.

Loftman, P. and Nevin, B. (1996) 'Going for growth: Prestige projects in three British cities', *Urban Studies* 33, (6), pages 991–1019.

Logie, G. (1962) 'The lessons of Coventry', *Architect and Building News* 11 July, pages 43–49.

Lomholt, L. (2016) *Coventry City Centre Masterplan* (available at: www.e-architect. co.uk/birmingham/coventry-city-centre-masterplan).

Low, K. (2013) 'Olfactive frames of remembering: Theorizing self, senses and society', *The Sociological Review* 61, pages 688–708.

Lowe, R. (1990) 'World War Two, consensus and the foundation of the welfare state', in *Twentieth Century British History* 1, pages 2–28.

Luder, O. (1964) 'Birmingham's Bull Ring', *Architect and Building News* 28 Aug, pages 400–410.

Madin, C. (2011) *John Madin, Architect and Planner: An Illustrated Record* (available at: www.john-madin.info/downloads/John_Madin_Architect_Planner_ebook.pdf).

Mah, A. (2012) 'Demolition for development: A critical analysis of official urban imaginaries in past and present UK cities', *Journal of Historical Sociology* 25, (1), pages 151–176.

Malpass, P. (2003) 'The Wobbly Pillar? Housing and the British postwar welfare state', *Journal of Social Policy* 32, (4), pages 589–606.

Mandler, P. (1999) 'New towns for old: The fate of the town centre', in Conekin, B., Mort, F., and Waters, C. (eds.) *Moments of Modernity: Reconstructing Britain, 1945–1964* Rivers Oram Press, London.

Manzoni, H.J. (1942) 'Post-war planning and reconstruction', *Institution of Civil Engineers Journal* 18 June, paper 5317, pages 233–254, and discussion, *ICE Journal* 18 Oct, pages 534–557. This paper was awarded a Telford Premium by the Institution.

Manzoni, H.J. (1943) 'Duddeston and Nechells redevelopment area', *Report to the Public Works Committee* Birmingham City Council, Birmingham, Copy in Birmingham Central Reference Library.

Manzoni, H.J. (1955) 'Redevelopment of blighted areas in Birmingham', *Journal of the Town Planning Institute* Mar, pages 90–102.

Manzoni, H.J. (1961) 'The inner ring road, Birmingham', *Proceedings of the Institute of Civil Engineers* 18, Mar, pages 265–283.

Manzoni, H.J. (1968) 'The development of town planning in Birmingham', Notes from the History of Birmingham Seminar no. 2, School of History, University of Birmingham, Birmingham City Council, Birmingham.

Marriott, O. (1967) *The Property Boom* H. Hamilton, London.

Marwick, A. (1968) *Britain in the Century of Total War: War, Peace and Social Change 1900–1967* Penguin, London.

Matless, D. (1998) *Landscape and Englishness* Reaktion Books, London.

McEwan, C., Pollard, J. and Henry, N. (2005) 'The "global" in the city economy: Multicultural economic development in Birmingham', *International Journal of Urban and Regional Studies* 29, pages 916–933.

McGrory, D. (2015) *Coventry's Blitz* Amberley Publishing, Stroud, Gloucestershire.

McKenna, J. (2005) *Birmingham: The Building of a City* Tempus, Stroud.

Merriman, P. (2008) *Driving Spaces: A Cultural-Historical Geography of England's M1 Motorway* Wiley, London.

Midland Daily Telegraph (1939) 'Letter', 25 Feb.

Miles, M. (2002) 'Wish you were here', in Spier, S. (ed.) *Urban Visions: Experiencing and Envisioning the City* Tate Liverpool Press, Liverpool.

Miller, D. (1989) *Lewis Mumford: A Life* Weidenfeld and Nicholson, New York.

Millet, L. (2000) 'Coventry: Making a better future', in Falk, N. (ed.) *The Renaissance of Post-War Town Centres* URBED, London, page 38.

Moore, R. (2011) 'Park Hill estate, Sheffield: Review', *The Guardian* 20 Aug.

Moran, J. (2010) *On Roads: A Hidden History* Profile Books Ltd., London.

MoWT (1946) *Design and Layout of Roads in Built-Up Areas* HMSO, London.

MTCP (1947) *Advisory Handbook on the Redevelopment of Central Areas* HMSO, London.

Mullen, C. (2018) 'fDi's European Cities and Regions of the Future 2018/19 – Cities', https://www.fdiintelligence.com/Locations/Europe/fDi-s-European-Cities-and-Regions-of-the-Future-2018-19-Cities.

Mumford, L. (1938) *The Culture of Cities* Harcourt Brace Jovanowich, New York.

Muzaini, H. (2015) 'On the matter of forgetting and "memory returns"', *Transactions of the Institute of British Geographers* 40, (1), pages 102–112.

Myles Wright, H. (1955) 'The first ten years: Post-war planning and development in England', *Town Planning Review* 26 July, pages 73–91.

Nairn, I. (1960) 'Birmingham: Liverpool: Manchester', *Architectural Review* 127, (762), pages 111–116.

Nasr, J. and Voilait, M. (2003) *Urbanism: Imported or Exported?* Wiley, Chichester, Sussex.

Nathaniel Lichfield and Partners (2008) *Coventry Retail Study* Prepared for Coventry City Council, Coventry.

Newbold, E.B. (1982) *Portrait of Coventry* Hale, London.

New Economics Foundation (2010) *Re-Imagining the High Street: Escape from Clone Town Britain, the 2010 Clone Town Report* NEF, London.

Nicholas, R. (1945) *City of Manchester Plan: Prepared for the City Council* Jarrold, Manchester.

Nozlopy, G. (2003) *Public Sculpture of Warwickshire, Coventry and Solihull* Liverpool University Press, Liverpool.

Oc, T. and Tiesdell, S. (1998) 'City centre management and safer city centres: Approaches in Coventry and Nottingham', *Cities* 15, (2), pages 85–103.

Orazi, S. (2015) *Modernist Estates: The Buildings and the People Who Live in Them* Frances Lincoln Limited, London.

Orwell, G. (1938) *Homage to Catalonia* Secker & Warburg, London.

Orwell, G. (1939) *Coming Up for Air* Penguin, London.

Osborn, F. (1942) *The Land and Planning: Elements of the Problems of Compensation and Betterment and Land Values* Faber and Faber, London.

Parker, D. and Long, P. (2000) 'Reimagining Birmingham: Public history, selective memory and the narration of urban change', *European Journal of Cultural Studies* 6, pages 157–178.

Parker, D. and Long, P. (2004) '"The mistakes of the past"? Visual narratives of urban decline and regeneration', *Visual Culture in Britain* 5, pages 37–58.

Pendlebury, J. (2003) 'Planning the historic city: 1940s reconstruction plans in Britain', *Town Planning Review* 74, (4), pages 371–393.

Pendlebury, J. (2004) 'Reconciling history with modernity: 1940s plans for Durham and Warwick', *Environment and Planning B: Planning and Design* 31, (3), pages 331–348.

Pendlebury, J. (2009) 'The urbanism of Thomas Sharp', *Planning Perspectives* 24, (1), pages 3–27.

Perkins, C. and Dodge, M. (2012) 'Mapping the imagined future: The roles of visual representation in the 1945 city of Manchester plan', *Bulletin of the John Rylands Library* 89, (1), pages 247–276.

Petty, J. (1987) *Coventry Cathedral: After the Flames* Jarrold Colour Publications, Coventry.

Pevsner, N. (1951) 'Canons of criticism', *Architectural Review* Jan, pages 3–6.

Pickford, N. and Pevsner, N. (2016) *Warwickshire: The Buildings of England* Penguin, London.

Platt, H.L. (2015) *Building the Urban Environment: Visions of the Organic City in the United States, Europe, and Latin America (Urban Life, Landscape and Policy)* Temple University Press, Philadelphia, US.

Pollard, J. (2004) 'From industrial district to "urban village"? Manufacturing, money and consumption in Birmingham's jewellery quarter', *Urban Studies* 41, (1), pages 173–193.

Portas, M. (2011) *The Portas Review: An Independent Review into the Future of Our High Streets* Department for Business, Innovation and Skills, London.

Price, F. (1960) 'Making a clean sweep to a city of the future', in *The New Birmingham* Mail, Birmingham.

Price, F. (2002) *Being There* Upfront, Kibworth.

Prince of Wales (1989) *A Vision of Britain: A Personal View of Architecture* Doubleday, London.

Pugh, A.R. and Percy, A.L. (1946) 'Planning and reconstruction in Coventry', *Journal of the Institution of Municipal and County Engineers* 30 Sept, pages 73–92.

Punter, J.V. (1990) *Design Control in Bristol 1940–1990* Redcliffe Press, Bristol.

Punter, J.V. (1994) 'Design control in England', *Built Environment* 20, pages 169–180.

Raco, M. (2003) 'Remaking place and securing space: Urban regeneration and the strategies, tactics and practices of policing in the UK', *Urban Studies* 40, (9), pages 1869–1887.

Ray, J. (1996) *The Night Blitz 1940–1941* Arms and Armour Press, London.

Redknap, B. (2004) *Engineering the Development of Coventry* Branch of the Historical Association, Coventry.

Reith, J.S.W. (1949) *Into the Wind* Hodder & Stoughton, London.

Relph, E. (1976) *Place and Placelessness* Pion, London.

Richards, J. (1952) 'Coventry', *Architectural Review* Jan, pages 3–7.

Richardson, K. (1972) *Twentieth Century Coventry* City of Coventry, Coventry.

Rigby-Childs, D. (1954) 'A comparison of progress in rebuilding bombed cities', *Architects' Journal* 8 July, pages 41–42.

Rigby-Childs, D. and Boyne, D. (1953) 'Coventry', *Architect's Journal* 107, 8 Oct, pages 428–447.

Rigby-Childs, D. and Boyne, D.A. (1953) 'A survey of Coventry, with regard to replanning and rebuilding bombed areas, and the town as it now is', *Architects' Journal* 8 Oct, pages 429–434.

Rothnie, N. (1992) *The Baedeker Blitz* Ian Allen, Shepperton.

Saarinen, E. (1943) *The City: Its Growth–Its Decay–Its Future* Reinhold, New York.

Saint, A. (2004) 'Gibson, Sir Donald Edward Evelyn (1908–1991)', in *Oxford Dictionary of National Biography* Oxford University Press, Oxford.

Sandercock, L. (1998) *Making the Invisible Visible: A Multicultural Planning History* University of California Press, Berkeley and London.

Sandercock, L. (2003) *Cosmopolis II: Mongrel Cities of the 21st Century* Continuum, London and New York.

Scott, J. (1998) *Seeing Like a State: How Certain Schemes to Improve the Human Condition Have Failed* Yale University Press, Yale.

Sennett, R. (1970 [1990]) *The Uses of Disorder* Faber and Faber, London.

Shapeley, P. (2012) 'Civic pride and policy in the post-war city', *Urban History* 39, (2), pages 310–328.

Sharp, T. (1936 [1938]) *English Panorama* Dent, London.

Sharp, T. (1940) *Town Planning* Pelican, Harmondsworth.

Sharp, T. (1946) *Exeter Phoenix: A Plan for Rebuilding* Architectural Press, London.

Sharp, T. (1948) *Oxford Replanned* Architectural Press, London.

Sharp, T. (1949) *A Plan for Salisbury* Architectural Press, London.

Sharp, T. (1952) 'Building Britain: 1941', *Town Planning Review* 23, pages 206–210.

Sheail, J. (1981) 'Sir Graham Vincent: An Appreciation', *Planning History Bulletin* 3, (3), pages 13–17.

Sheail, J. (2012) 'British inter-war planning: The recollections of a government official', *Planning Perspectives* 27, (2), pages 285–296.

Short, J.R. (2014) *Urban Theory: A Critical Assessment* (2nd edition) Palgrave Macmillan, New York.

Sibley, D. (1995) *Geographies of Exclusion: Society and Difference in the West* Routledge, London.

Skipp, V. (1980) *A History of Greater Birmingham Down to 1830* Privately Published, Birmingham.

Smith, O.S. (2015) 'Central government and town-centre redevelopment in Britain, 1959–1966', *The Historical Journal* 58, (1), pages 217–244.

Spence, B. (1962) *Phoenix at Coventry: The Building of a Cathedral* Harper and Row, London.

Spring, M. (2003) 'The fellowship of the bullring', *Building* 34, pages 1–15.

Stansfield, K. (1981) 'Thomas Sharp 1901–1978', in Cherry, G.E. (ed.) *Pioneers in British Town Planning* Architectural Press, London.

Steadman, M. (1968) 'Birmingham builds a model city', *Geographical Magazine* 40, (16), page 1383.

Stephens, W.B. (ed.) (1969) 'The city of Coventry: Buildings: Domestic buildings', in *A History of the County of Warwick: Volume 8: The City of Coventry and Borough of Warwick* Institute of Historical Research, Coventry.

Stevenson, J. (1986) 'Planner's moon? The Second World War and the planning movement', in Smith, H. (ed.) *War and Social Change* Manchester University Press, Manchester, pages 58–77.

Stevenson, J. (1991) 'The Jerusalem that failed? The rebuilding of post-war Britain', in Gourvish, T. and O'Day, A. (eds.) *Britain Since 1945: Problems in Focus Series* Palgrave, London.

Sutcliffe, A.R. (1967–1969) 'Transcripts of interviews with H.J. Black, Alderman T.W. Bowen, J. Lewis, J. Madin, Sir Herbert Manzoni, Sir Frank Price, R.F.H. Ross', in *Transcripts with Prominent Birmingham People 1967–1969* Birmingham Central Library.

Sutcliffe, A.R. (1975) 'Case studies in modern British planning history: The Birmingham Inner Ring Road', paper submitted for discussion at the History of Planning Group meeting, Birmingham (copy in Ginsburg collection, University archives, Birmingham City University).

Sutcliffe, A.R. and Smith, R. (1974) *History of Birmingham 1939–1970* Oxford University Press, Oxford.

Tait, M. and While, A. (2009) 'Ontology and the Conservation of Built Heritage', *Environment and Planning D: Society and Space* 27, (4), pages 721–737.

Tallon, A. (2010) *Urban Regeneration and Renewal* Routledge, London.

Taylor, A.J.P. (1961) *The Origins of the Second World War* Penguin, London.

Taylor, N. (1969) 'Foreword', in Lewison, G. and Billingham, R. (eds.) *Coventry New Architecture* Edward, *The* Printers Ltd, Coventry.

Taylor, F. (2015) *Coventry: November 14, 1940* Bloomsbury, US.

TCPA (1943) *Rebuilding Britain*, A select list of books on town and country Planning, Bristol Public Library, Bristol.

Tewdwr-Jones, M. (2011) *Urban Reflections* Policy Press, University of Bristol.

Tewdwr-Jones, M. (2013) 'From Town Hall to Cinema: Documentary film as planning propaganda in post-war Britain', *Berkeley Planning Journal* 26, pages 86–106.

Tibbalds, F., Colbourne, Karski, Williams and Monro (1990) *Birmingham Urban Design Strategy* Birmingham City Council, Birmingham.

Till, K. (2012) 'Wounded cities: Memory-work and a place-based ethics of care', *Political Geography* 31, pages 3–14.

Tiptaft, N. (1947) *So This Is Birmingham* Norman Tiptaft Ltd, Birmingham.

Tiratsoo, N. (1990) *Reconstruction, Affluence and Labour Politics: Coventry, 1945–1960* Routledge, London.

Tiratsoo, N., Hasegawa, J., Mason, T. and Matsumura, T. (2002) *Urban Reconstruction in Britain and Japan, 1945–1955: Dreams, Plans and Realities* University of Luton Press, Luton.

Titmus, R. (1950) *Problems of Social Policy* HMSO, London.

Tripp, H.A. (1938) *Road Traffic and Its Control* Edward Arnold, London.

Tripp, H.A. (1942) *Town Planning and Road Traffic* Edward Arnold, London.

Tubbs, R. (1942) *Living in Cities* Penguin, London.

Upton, C. (1993) *A History of Birmingham* Phillimore, Chichester.

van Nes, E. (2001) 'Road building and urban change', *Proceedings 3rd International Space Syntax Symposium*, Atlanta.

Voldman, D. (1990) 'Reconstructors' tales: An example of the use of oral sources in the history of reconstruction after the Second World War', in Diefendorf, J. (ed.) *Rebuilding Europe's Bombed Cities* Macmillan, Basingstoke, pages 238–245.

Walford, S. (2009) 'Architecture in tension: An examination of the position of the architect in the private and public sectors, focusing on the training and careers of Sir Basil Spence (1907–1976) and Sir Donald Gibson (1908–1991)' Unpublished PhD, Department of the History of Art, University of Warwick.

Walters, P. (2014) *The Story of Coventry* History Press, Stroud, Gloucestershire.

Ward, S. (1999) 'The international diffusion of planning: A review and a Canadian case study', *International Planning Studies* 4, (1), pages 53–77.

Ward, S. (2004) *Planning and Urban Change* Paul Chapman, London.

Ward, S.V. (2012) 'Gordon Stephenson and the "galaxy of talent": Planning for postwar reconstruction in Britain 1942–1947', *Town Planning Review* 83, (3), pages 279–296.

Ward, S.V. (2017) 'Planning diffusion: Agents, mechanisms, networks, and theories', in Hein, C. (ed.) *The Routledge Handbook of Planning History* Routledge, London, pages 90–104.

Ward, S.V., Freestone, R. and Silver, C. (2011) 'The "new" planning history: Reflections, issues and directions', *Town Planning Review* 82, (3), pages 231–262.

Watson, J. and Abercrombie, P. (1943) *A Plan for Plymouth* Underhill, Plymouth.

West Midlands Group (1948) *Conurbation: A Survey of Birmingham and the Black Country* Architectural Press, London.

Wheatcroft, S. (2013) 'Holiday camps, castles and stately homes: The residential option for the evacuation of disabled children during World War II', in Clapson, M. and Larkham, P.J. (eds.) *The Blitz and Its Legacy: From Destruction to Reconstruction* Ashgate, Aldershot.

While, A. (2006) 'Modernism vs urban renaissance: Negotiating post-war heritage in English city-centres', *Urban Studies* 43, pages 2399–2419.

While, A. and Pendlebury, J. (2008) 'Modern Movement conservation as progressive practice: Byker and British welfare state housing', in Van Den Heuvel, D., Mesman, M., Quist, W. and Lemmens, B. (eds.) *Proceedings of the 10th International DOCOMOMO Conference* Delft University Press, Delft.

While, A. and Short, M. (2011) 'Place narratives and heritage management: The modernist legacy in Manchester', *Area* 43, (1), pages 4–13.

Wildavsky, A. (1973) 'If planning is everything, maybe it's nothing', *Policy Sciences* 4, (2), pages 127–153.

Williams, J. (2017) *Shearer Property Group appointed for major Coventry development* (available at: https://developmentfinancetoday.co.uk/article-desc-5094_shearer-property-group-appointed-for-major-coventry-development).

Yelling, J.A. (1986) *Slums and Slum Clearance in Victorian London* Alfred & Unwin, London.

Yiftachel, O. (1998) 'Planning and social control: Exploring the dark side', *Journal of Planning Literature* 12, (4), pages 395–406.

Yoneyama, L. (1999) *Hiroshima Traces: Time, Space, and the Dialectics of Memory* University of California Press, Berkeley and Los Angeles, CA.

Young, P. (1966) *World War 1939–1945* Barker, London.

Zukin, S. (1998) 'Urban lifestyles: Diversity and standardization in spaces of consumption', *Urban Studies* 35, (5–6), pages 825–839.

Zukin, S. (2010) *Naked City* Oxford University Press, New York.

Index

Note: Page numbers in italics refer to figures on the corresponding pages.

Printed in the United States
by Baker & Taylor Publisher Services